# Techniques in Life Science and Biomedicine for the Non-Expert

**Series Editor**
Alexander E. Kalyuzhny, University of Minnesota
Minneapolis, MN, USA

The goal of this series is to provide concise but thorough introductory guides to various scientific techniques, aimed at both the non-expert researcher and novice scientist. Each book will highlight the advantages and limitations of the technique being covered, identify the experiments to which the technique is best suited, and include numerous figures to help better illustrate and explain the technique to the reader. Currently, there is an abundance of books and journals offering various scientific techniques to experts, but these resources, written in technical scientific jargon, can be difficult for the non-expert, whether an experienced scientist from a different discipline or a new researcher, to understand and follow. These techniques, however, may in fact be quite useful to the non-expert due to the interdisciplinary nature of numerous disciplines, and the lack of sufficient comprehensible guides to such techniques can and does slow down research and lead to employing inadequate techniques, resulting in inaccurate data. This series sets out to fill the gap in this much needed scientific resource.

More information about this series at https://link.springer.com/bookseries/13601

Akash Gautam

# DNA and RNA Isolation Techniques for Non-Experts

 Springer

Akash Gautam (iD)
Center for Neural and Cognitive Sciences
University of Hyderabad
Hyderabad, Telangana, India

ISSN 2367-1114                    ISSN 2367-1122    (electronic)
Techniques in Life Science and Biomedicine for the Non-Expert
ISBN 978-3-030-94232-8            ISBN 978-3-030-94230-4    (eBook)
https://doi.org/10.1007/978-3-030-94230-4

This Springer imprint is published by the registered company Springer Nature Switzerland AG
The registered company address is: Gewerbestrasse 11, 6330 Cham, Switzerland

# Sources and Acknowledgments

The arduous task of compiling all the latest DNA and RNA isolation techniques could not have been possible without the help of my colleagues and experts from different fields. I am highly indebted to them for their constant support and agreeing to be co-authors for the following chapters in this book.

I am first and foremost thankful to my friend Anisha David (Ph.D. in Botany) for her contribution to Chap. 1 (Basic Idea About Nucleic Acids for Nonexperts) and Chap. 13 (CTAB- or SDS-Based Isolation of Plant's DNA). Her remarkable zeal and enthusiasm in writing both chapters despite suffering from Covid-19 are highly appreciable.

I am grateful to Vishal Chaudhary, Ph.D. in Physics with a command in nanomaterials, for co-authoring Chap. 5 (Spin Column-Based Isolation of Nucleic Acid) and Chap. 15 (Magnetic Bead-Based Nucleic Acid Isolation). Both chapters are based on the biophysical properties of isolating materials, and he has done absolute justice with both chapters by meticulously explaining the principle behind these techniques.

Dr. Ajeet Kumar Singh and Ms. Priya Agarwal, both from the Department of Zoology, CMP College (Prayagraj), have helped me in writing Chap. 7 (DNA Isolation by Hydrophilic Ionic Liquid Treatment), Chap. 9 (Isolation of Bacteriophage DNA by PEG Method), and Chap. 10 (DNA Isolation by Chelex Method). I am thankful to both of them for making these chapters more comprehensible for readers from diverse backgrounds.

Many thanks to Drs. Neelabh and Prerna Giri duo for co-authoring Chap. 17 (DNA Extraction from Agarose Gel Through Paper Strips). Further, their support in writing Chap. 16 (Density Gradient-Based Nucleic Acid Isolation) along with Anupam Singh and Ashish Kumar Rai is highly appreciated. This whole team of biotechnologists has done an outstanding work on these chapters by writing them in scientific yet straightforward language.

I would not be able to get my work done without the continual support of my university juniors, Drs. Mukesh Meena and Prashant Swapnil. Both have co-authored Chap. 21 (Southern and Northern Blotting) conscientiously as well as

Chap. 18 (Transformation or Genetic Modification of Cells/Organism) along with Rahul Kumar.

I would like to acknowledge the extraordinary debt I owe to Dr. Rohini Motwani (an MBBS/MS doctor) and Dr. Rohit Saluja (a medical science researcher) for contributing their knowledge for Chap. 12 (Isolation of DNA from Blood Samples by Salting Method) along with their team comprising Himadri Singh, Manisha Naithani, Ashok Kumar, Mirza S Baig, and Kiran Kumar. Dr. Motwani and Dr. Saluja have also contributed to two applications-based book chapters: Chap. 19 (Gene Cloning and Vectors) along with Anupam Jyoti and Juhi Saxena and Chap. 23 (Applications of DNA Sequencing Technologies for Current Research) together with Rajeev Nema, Avinash Narayan, and Ashok Kumar.

Finally, I would like to acknowledge with gratitude the support and love of my family and my students, who all kept me going, and this book would not have been possible without them.

# Contents

# Contributors

**Priya Agrawal** Department of Zoology, CMP Degree College, University of Allahabad, Prayagraj, India

**Mirza S. Baig** Discipline of Biosciences and Biomedical Engineering (BSBE), Indian Institute of Technology, Simrol, Indore, India

**Vishal Chaudhary** Department of Physics, Bhagini Nivedita College, University of Delhi, New Delhi, India

**Anisha David** Bengaluru, India

**Akash Gautam** Center for Neural and Cognitive Sciences, University of Hyderabad, Hyderabad, Telangana, India

**Prerna Giri** Department of Zoology, Banaras Hindu University, Varanasi, India

**Anupam Jyoti** Faculty of Applied Sciences and Biotechnology, Shoolini University of Biotechnology and Management Sciences, Bajhol, Solan, India

**Ashok Kumar** Department of Biochemistry, All India Institute of Medical Sciences, Saket Nagar, Bhopal, India

**Kiran Kumar** Department of Biotechnology & Bioinformatics, FLS, JSS Academy of Higher Education & Research, Mysuru, India

**Rahul Kumar** International Centre for Genetic Engineering and Biotechnology, New Delhi, India

**Mukesh Meena** Department of Botany, Mohanlal Sukhadia University, Udaipur, India

**Rohini Motwani** Department of Anatomy, All India Institute of Medical Sciences, Bibinagar, Hyderabad, India

**Manisha Naithani** Department of Biochemistry, All India Institute of Medical Sciences, Rishikesh, India

**Avinash Narayan** Department of Biochemistry, All India Institute of Medical Sciences, Saket Nagar, Bhopal, India

**Neelabh** Department of Biotechnology, Bansal Institute of Engineering and Technology, Lucknow, India

**Rajeev Nema** Department of Biochemistry, All India Institute of Medical Sciences, Saket Nagar, Bhopal, India

**Ashish Kumar Rai** Department of Biotechnology, Bansal Institute of Engineering and Technology, Lucknow, India

**Rohit Saluja** Department of Biochemistry, All India Institute of Medical Sciences, Bibinagar, Hyderabad, India

**Juhi Saxena** Faculty of Applied Sciences and Biotechnology, Shoolini University of Biotechnology and Management Sciences, Bajhol, Solan, India

**Ajeet Kumar Singh** Department of Zoology, CMP Degree College, University of Allahabad, Prayagraj, India

**Anupam Singh** Department of Biotechnology, Bansal Institute of Engineering and Technology, Lucknow, India

**Himadri Singh** Department of Biochemistry, All India Institute of Medical Sciences, Saket Nagar, Bhopal, India

**Prashant Swapnil** Department of Botany, School of Biological Science, Central University of Punjab, Ghudda, Bhatinda, India

# Part I
# Introductory Chapters

# Basic Idea About Nucleic Acids for Nonexperts

**Abstract** Nucleic acids form the basis of all life on earth. The two types of nucleic acids are DNA and RNA. DNA primarily stores genetic information, whereas RNA is mainly involved in the processing (transcription) and manifestation (translation) of this genetic information. They are unbranched macromolecules made up of monomeric units called nucleotides which are linked together by phosphodiester bonds. Nucleotides consist of a pentose sugar, a nitrogenous base, and a phosphate group. Two antiparallel strings of nucleotides coil around a central axis forming the double-helical structure of DNA, which in association with proteins, mainly the histones, undergoes several degrees of folding and gives rise to chromatin. Whereas RNA is primarily single-stranded and depending upon the role it plays, it may be genetic or non-genetic. The non-genetic RNAs like mRNA, rRNA, tRNA, snRNA, and miRNA play various functional roles. Apart from being a storehouse of genetic information, nucleic acids also undergo mutations that lead to the expression of new traits and serve as the foundation for evolution.

**Keywords** Nucleic acids · DNA double helix · Types of DNA · Chromatin · mRNA · rRNA · tRNA

## 1 Molecules of Life

Life began with the formation of the first cell in the primitive earth about 3.8 billion years ago. The environmental conditions prevailing in the primitive earth led to the formation of various organic molecules like proteins, carbohydrates, lipids, and nucleic acids. These molecules aggregated, got enclosed in a phospholipid membrane, and began orchestrating the biological machinery over time. These primitive cells eventually started synthesizing all the molecules of life. These cells thrived and created progenies, not missing any information about the cellular design, concoction, and machinery (Fig. 1). This information with absolute detailing was secured in the nucleic acids and transmitted faithfully to the next generation. From thereon,

© The Author(s), under exclusive license to Springer Nature Switzerland AG 2022
A. Gautam, *DNA and RNA Isolation Techniques for Non-Experts*, Techniques in Life Science and Biomedicine for the Non-Expert,
https://doi.org/10.1007/978-3-030-94230-4_1

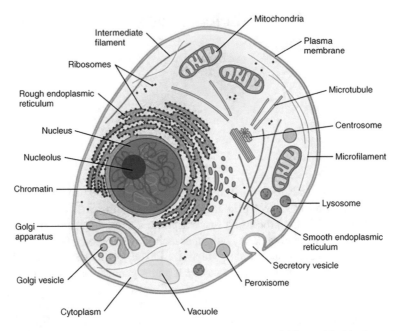

**Fig. 1** Structure of an animal cell, displaying the genetic material housed inside the nucleus. OpenStax, CC BY 4.0 https://creativecommons.org/licenses/by/4.0, via Wikimedia Commons. (Source: imgbin.com)

life continued and evolved, bringing forth tremendous diversity. Undoubtedly, nucleic acids found in all living organisms are pivotal to life.

## 2  Nucleic Acids

Nucleic acids are present as deoxyribonucleic acid (DNA) and ribonucleic acid (RNA) in all cells. DNA primarily stores the genetic information, but in most plant viruses and some animal viruses, RNA serves as the genetic material. Nucleic acids are unbranched macromolecules made up of monomeric units called nucleotides consisting of a pentose sugar, a nitrogenous base, and a phosphate group (Fig. 2). The nucleotides of DNA are called deoxyribonucleotides, and those of RNA are called ribonucleotides.

*Pentose* is a monosaccharide with five carbon atoms, which exist as furanose (five-membered ring) in nucleic acids. The carbon atoms are numbered with primes to distinguish them from the atoms of the neighboring nitrogenous bases. RNA contains the pentose sugar ribose, whereas DNA contains deoxyribose (Fig. 3). The two sugars differ in their 2′ carbon atoms, with ribose ($C_5H_{10}O_5$) having a hydroxyl

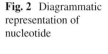

**Fig. 2** Diagrammatic representation of nucleotide

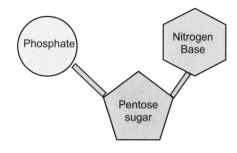

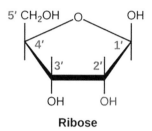
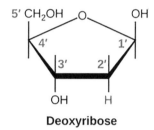

**Fig. 3** The pentose sugars. (Source: Figure_09_01_04f.jpg)

(–OH) group and deoxyribose ($C_5H_{10}O_4$) having only hydrogen (H). It is the deoxyribose moiety with which Feulgen stain specifically reacts and serves as a DNA stain. Hence, deoxyribose has one oxygen atom less than ribose, making the DNA less reactive, more stable, and better suited for long-term storage of genetic information (Pierce, 2005). Thus, gradually during evolution, RNA, which preceded DNA as the genetic material, gave way to the DNA world (Watson et al., 2004).

*Nitrogenous bases* are heterocyclic compounds containing carbon and nitrogen in their rings. These compounds are weakly basic in nature and are therefore called bases. They are categorized as the bicyclic purines and the monocyclic pyrimidines. Purines have two fused rings, one of which is hexagonal, and the other is pentagonal. Adenine (**A**, $C_5H_5N_5$) and guanine (**G**, $C_5H_5ON_5$) constitute the purines (Fig. 4). Pyrimidines have a single hexagonal ring consisting of cytosine (**C**, $C_4H_5ON_3$), thymine (**T**, $C_5H_6O_2N_2$), and uracil (**U**, $C_4H_4O_2N_2$) (Fig. 5). Uracil is found only in RNA, whereas thymine is restricted to DNA; this difference has enabled cell biologists to use their radioactive forms as specific labels for RNA and DNA, respectively.

As heterocyclic bases absorb UV at a wavelength of 260 nm, cells photographed in this wavelength show nucleolus, chromatin, and cytoplasmic regions containing RNA (Fig. 6). The concentration of nucleic acids in a solution can also be estimated by measuring the absorption at 260 nm (Robertis & Robertis, 2006).

In a nucleotide, the nitrogenous base bonds with the 1′ carbon of the pentose sugar by an N-glycosidic bond, and this entity (sugar + base) is called a nucleoside. The following table provides an overview of different nucleosides and nucleotides.

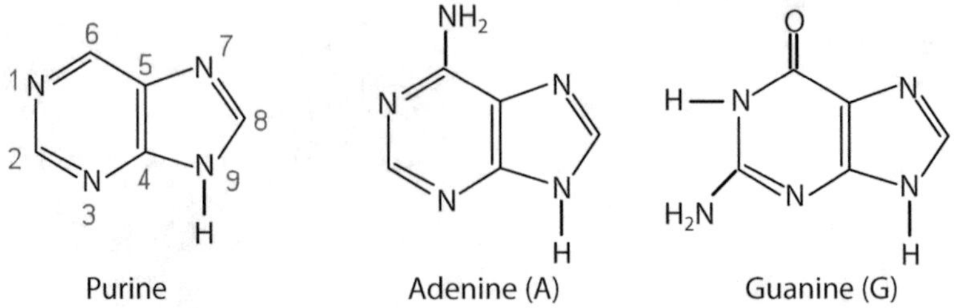

Purine                     Adenine (A)                    Guanine (G)

**Fig. 4** Purines: the bicyclic nitrogenous bases. (Source: Dc9df397a1d1801eb3c21a8630c64325.jpg)

Pyrimidine          Uracil (U)          Thymine (T)          Cytosine (C)
                    RNA only            DNA only             both DNA and RNA

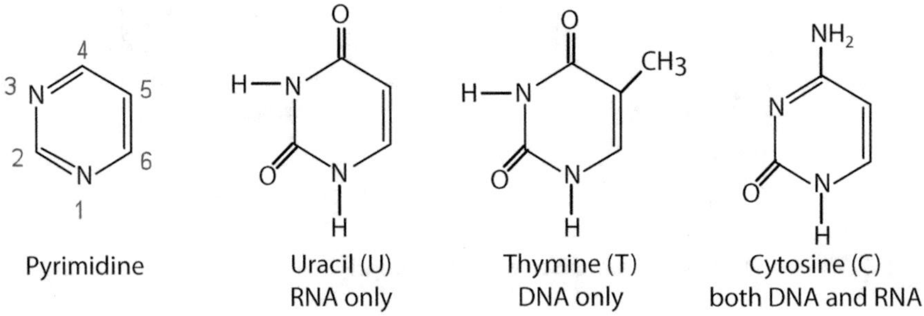

**Fig.     5** Pyrimidines:     the     monocyclic     nitrogenous     bases.     (Source: Dc9df397a1d1801eb3c21a8630c64325.jpg)

**Fig. 6** Neutrophil photographed at 260 nm (scale bar—5 μm), showing the lobed nucleus connected by chromatin. "Neutrophil" by Dr Franklund is licensed under CC BY 2.0. (Ojaghi et al., 2020)

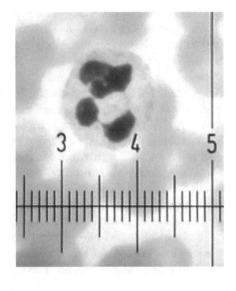

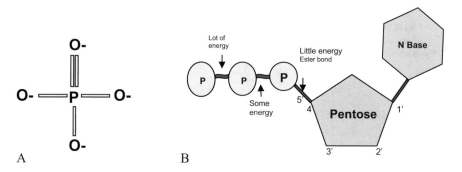

**Fig. 7** (**a**) Phosphate group, (**b**) diagrammatic representation of nucleotide triphosphate, showing the high-energy phosphate bonds

| Nitrogenous base | Nucleoside (RNA) | Nucleoside (DNA) | Nucleotide (RNA) | Nucleotide (DNA) |
|---|---|---|---|---|
| Adenine | Adenosine | Deoxyadenosine | Adenylate | Deoxyadenylate |
| Guanine | Guanosine | Deoxyguanosine | Guanylate | Deoxyguanylate |
| Cytosine | Cytidine | Deoxycytidine | Cytidylate | Deoxycytidylate |
| Thymine | Thymidine | Deoxythymidine | Uridylate | Deoxyuridylate |
| Uracil | Uridine | Deoxyuridine | Thymidylate | Deoxythymidylate |

*Phosphate group(s)* consists of a pentavalent phosphorus atom carrying four oxygen atoms (three bonded by a single bond and one by a double bond) (Fig. 7a).

A nucleotide may consist of one, two, or three phosphates linked to the 5′ carbon (of pentose) of a nucleoside by an ester bond-forming nucleoside monophosphate, diphosphate, or triphosphate, respectively. Nucleoside triphosphate serves as the substrate for nucleic acid synthesis and links with its 3′ carbon (of pentose) to the free 5′ end of the growing nucleotide chain. This free 5′ end still has three phosphates, and the energy released from the removal of two end phosphates leads to the formation of an ester bond with the 3′ carbon of the new nucleotide. Thus, the two nucleotides get connected utilizing a phosphodiester bond (Fig. 7b).

The phosphodiester bond involves two oxygen atoms of the phosphate group while the third oxygen ion remains free, conferring a net negative charge. This gives rise to the acidic properties of nucleic acid and enables them to bind to basic proteins called histones. The alternating phosphate and pentose sugar form the backbone of the polynucleotide chain, and the nitrogenous bases are attached almost perpendicular to this backbone (Pierce, 2005). Polynucleotide chains exhibit polarity, having a 5′ end containing a free triphosphate and the 3′ end, which is usually a free hydroxyl group (Fig. 8). Nucleoside triphosphates like ATP (adenosine triphosphate), in addition to serving as a substrate for nucleic acid synthesis, also store and transfer energy in their high-energy phosphate bonds. At the same time, other nucleotides play an essential role in cell signaling as messengers.

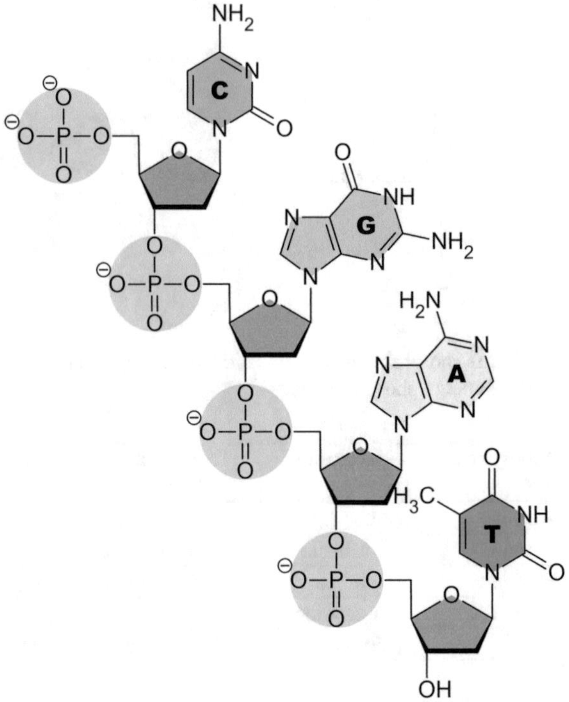

**Fig. 8** The primary structure of nucleic acid (DNA). (Source: imgbin.com)

## 3 DNA

DNA is a complex molecule that primarily serves as the hereditary material and has all information for the formation and maintenance of an organism. In many prokaryotes, DNA is present as a single circular supercoiled molecule associated with proteins forming a dense structure packaged in a region called a nucleoid. Some bacteria may also contain another smaller circular DNA molecule called a plasmid. On the contrary, eukaryotes have linear DNA molecules, which, in association with proteins, form condensed structures called chromosomes and are harbored inside the nucleus. Apart from the nuclear DNA, eukaryotes also contain small circular DNA in cell organelles like mitochondria and chloroplast.

DNA was first observed by Swiss chemist Friedrich Miescher in 1869 inside the nucleus of human white blood cells and called it nuclein. Following this discovery, many notable scientists performed a series of research and unraveled details about this molecule of life. All this work formed the foundation of the historical double-helix model of DNA proposed by Watson and Crick in 1953 (Pray, 2008). This DNA carries the genetic information in a linear sequence of the four base pairs (A,T,C,G). This four-letter alphabet codes for the primary structure of all proteins through a series of events and manifests itself. The genetic information stored in the DNA is passed on to RNA during transcription (the process of RNA synthesis). The nucleotide sequence of RNA contains the code for the amino acid sequence, which leads to the formation of proteins by translation. This pathway for the flow of genetic information is referred to as *central dogma* (Berg et al., 2002).

## 3.1 Secondary Structure of DNA: The Double Helix

In the year 1950, Erwin Chargaff published that the amount of A and T in DNA was almost the same, and so was the amount of C and G. Consequently, he proposed that DNA from any organism should have a 1:1 ratio of purines and pyrimidines. This regularity in base composition was explained by Watson and Crick's double-helix model of DNA. The 3-D structure proposed by Watson and Crick is that of B-DNA, which exists when abundant water surrounds the molecule and no unusual sequences are present. It is the most stable configuration, and evidence suggests that it is pre-dominant in the cellular environment (Pray, 2008).

B-DNA is a right-handed spiral structure or α-helix consisting of two polynucle-otide chains helically coiled around a central axis in a clockwise direction. This means, on climbing a spiral staircase, the outer railing will be in the right hand. These railings are made of negatively charged sugar and phosphate backbone, which runs antiparallel, i.e., the 5′-P end of one strand is close to the 3′-OH end of the other strand and vice versa. The diameter of this helix or the distance between sugar-phosphate backbones is 2 nm, and the nitrogen bases are stacked here as flat discs perpendicular to the helical axis. The nitrogen bases of the opposite strands pair by forming hydrogen bonds and holding the two nucleotide chains together. As the space between the sugar-phosphate backbones is just perfect for accommodat-ing a purine-pyrimidine pair, AT and GC are the only two possible base-pair combi-nations. A and T bind with two hydrogen bonds, while G and C bind with three, making the GC pair more stable than the AT pair (Fig. 9). The hydrophobic

**Fig. 9** The DNA double helix. Forluvoft, Public domain, via Wikimedia Commons. (Source: DNA-double-helix-768x990.jpg)

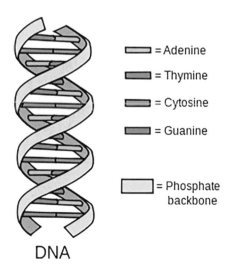

interactions between adjacent base pairs give further stability to the DNA helix (Ussery, 2002).

The specific base pairing ensures that if one strand has A, then the other strand will have T, and in the same way, if one strand has G, the other will have C. Hence, the two strands are not identical, but they are complementary, and if the base sequence of one strand is known, the sequence of the other strand can be determined. This property is of great significance during DNA replication, where the two chains separate, and both serve as a template for the synthesis of a new complementary chain. The resulting daughter helices have the same molecular constitution as the parent molecule.

The B-DNA has approximately ten bp per 360° rotation of the helix, and the helix twist per bp is 36°. It has a helix pitch (length of one complete turn of the helix) of 3.4 nm. Therefore, the helix rise per bp (distance between two adjacent bp) is 0.34 nm. The spiraling of nucleotide chains gives rise to minor and major grooves, which expose the hydrogen donors and acceptors of nitrogen bases and enable the binding of proteins to DNA (Watson et al., 2004).

## 3.2   Other Secondary Structures of DNA/Types of DNA

Though several different DNA double-helix structures have been studied, all have the basic pattern of two-helically coiled antiparallel chains. Depending on the humidity in the DNA's environment and the presence of unusual base sequences, DNA can assume other 3-D structures like A-DNA and Z-DNA.

A-DNA is found when the water content in the surrounding medium is comparatively low. It is an α-helix like the B-DNA but is relatively wider and more compressed with the base pairs tilting away from the helical axis. There is little evidence for the existence of this form of DNA in physiological conditions. This helix may be formed by RNA hybrids or DNA-RNA hybrids. It is also reported in the long terminal repeats (LTR) of transposable elements, which often contain purine stretches (Ussery, 2002).

Z-DNA is a left-handed helix where the sugar-phosphate backbone takes a zig-zag shape giving it the nomenclature. This is because the nitrogen bases, particularly the purines, are present in *syn* conformation and lie right on top of the sugar instead of the space between the backbone (this is in contrast to the *anti*-conformation of the bases found in B- and A-DNA). This form of DNA may appear in physiological conditions when particular base sequences are present, like a stretch of DNA with alternating C and G. This form is stable in synthetic double-stranded DNA with alternating purine-pyrimidine sequences (Turner et al., 2005). It is observed that parts of active DNA assume Z-DNA structure, suggesting its role in gene regulation (Pierce, 2005).

Some of the important contrasting features of different forms of DNA are given below in a tabular form (Pierce, 2005; Ussery, 2002).

| Parameter | B-DNA | A-DNA | Z-DNA |
|---|---|---|---|
| *Conditions required to produce the structure* | 92% water | 75% water | Alternating purine-pyrimidine bases |
| *Helical sense* | Right-handed | Right-handed | Left-handed |
| *Base pair per turn* | 10 | 11 | 12 (6 dimers) |
| *Helix pitch (nm)* | 3.4 | 2.8 | 4.5 |
| *Axial rise (nm)* | 0.34 | 0.26 | 0.37 |
| *Twist angle (°)* | 36 | 33 | −30 (−60 for dimer) |
| *Base-pair tilt (°)* | −6 | 20 | 7 |
| *Helix diameter (nm)* | 2 | 2.3 | 1.8 |
| *Overall shape* | Long and narrow | Short and wide | Elongated and narrow |

## 3.3 Physical and Chemical Properties of DNA

*Stability of nucleic acids*—Though hydrogen bonds enable base pairing and hold the two nucleotide chains together, the base stacking interactions between these base pairs confer overall stability to the helix. The flat aromatic nitrogen bases being hydrophobic tend to stack in an aqueous medium and twist slightly into an energetically favorable arrangement, giving rise to a helix. Thus, the hydrophobic interactions between the bases are the reason for the helical structure of DNA (Ussery, 2002).

*Viscosity*—DNA solutions are highly viscous because of their high axial ratio. DNA is thin and long, with a diameter of 2 nm and length ranging from a few micrometers to several centimeters. This very long length makes DNA molecules highly susceptible to damage by shearing forces or sonication (high-intensity ultrasound waves), leading to a considerable viscosity reduction.

*Spectroscopic properties*—As discussed earlier, nucleic acids absorb UV light and have a $\lambda_{max}$ at 260 nm. The absorbance is maximum for isolated nucleotides, intermediate for single-stranded DNA or RNA, and minimum for double-stranded DNA. This change in absorbance is termed hypochromicity, where double-stranded DNA is hypochromic compared to single-stranded DNA. The absorption properties of nucleic acids serve as excellent tools for detection, quantitation, and purity assessment (Berg et al., 2002).

*Effect of temperature*—The separation of the two strands of a double helix by breaking hydrogen bonds and hydrophobic interactions is known as *melting or denaturation*. An increase in temperature results in the denaturation of nucleic acids. The temperature at which 50% of the helix is denatured is known as the temperature of melting or the $T_m$ value of nucleic acid. $T_m$ value depends on the AT/GC ratio, as the GC pair needs more energy to break as compared to the AT pair. A 40% increase in absorbance accompanies the melting of the DNA helix (Watson et al., 2004).

The thermal denaturation of the double helix is reversible. If the denatured DNA is cooled slowly, the complementary bases pair by re-establishing hydrogen bonds and weak hydrophobic interactions, restoring the native conformation. Slow cooling allows time to find wholly complementary DNA strands, making the sample fully double-stranded. This process is termed renaturation or reannealing. On the other hand, rapid cooling allows base pairing and helix formation in local regions within or between DNA strands. DNA renaturation is a valuable tool in molecular biology as it can be used to estimate genome size by the time taken for reannealing. The longer the sequences, the greater the time taken for reannealing. DNA with repeat sequences (satellite DNA) renature faster (Robertis & Robertis, 2006).

*Effect of acid*—In strong acids (perchloric acid, $HClO_4$) and high temperatures (100 °C), nucleic acids are completely hydrolyzed into sugar, phosphate, and base. Slightly dilute mineral acids (pH—3–4) selectively break the glycosidic bonds with purines (easy to hydrolyze), resulting in apurinic nucleic acids. Advanced chemical reactions exist that selectively cleave nucleic acid strands and remove specific bases. This forms the basis for the DNA sequencing method developed by Maxam and Gilbert (Turner et al., 2005).

*Effect of alkali*—Increase of pH above the physiological range (pH 7–8) leads to the change in tautomeric states of the bases (keto to enol form). This affects the specific hydrogen bonding between the base pairs resulting in the denaturation of DNA by breaking the double-stranded structure. However, in RNA, an increase in pH leads to its hydrolysis by the cleavage of RNA backbone and formation of 2′,3′-cyclic phosphodiester bonds (Turner et al., 2005).

*Chemical denaturation*—Chemical agents like urea and formamide cause the denaturation of nucleic acids at neutral pH by destabilizing the energetics between the stacked hydrophobic bases.

## 3.4   From DNA to Chromosome

The DNA content is characteristic of each species and is referred to as *C-value*. Prokaryotes generally contain a single circular DNA, and their C-value is less than that of eukaryotes. *E. coli* DNA has $4.6 \times 10^6$ bp stretching to a total circumference of 1.5 mm. This long circular DNA is condensed and packaged into the tiny nucleoid by DNA supercoiling and the binding of various architectural proteins. Supercoiling can be imagined as twisting a rubber band such that it forms tiny coils and then twisting it further such that these tiny coils fold over each other to form a condensed ball. Most prokaryotic DNA is negatively supercoiled, which means the direction of the twist of DNA is opposite to the double helix. Multiple proteins act together to bring about the condensation of DNA by supercoiling, and one among them is the HU protein. Other proteins help to maintain this supercoiling (Griswold, 2008).

As compared to prokaryotes, higher organisms have multi-folds of DNA. Humans have about $3 \times 10^9$ bp per haploid genome, divided into 24 distinct DNA molecules called chromosomes (22 autosomes, x and y sex chromosomes), having a total length of 174 cm when fully extended. The length of DNA is measured in pico-grams (1 pg = $10^{-12}$ g), which is equivalent to 31 cm of DNA. Therefore, man has 3 pg of DNA per haploid genome. Chromosome 1, the largest one in the human genome, which contains 249 million bps, is 10 μm long in the metaphase stage, which means it gets compacted 7000 times. This means excellent packing and fold-ing is required to accommodate all this DNA inside the tiny nucleus. The extent of condensation changes through time, being elongated and relatively less condensed during the interphase. The extent of DNA packaging also changes locally during replication and transcription (Robertis & Robertis, 2006).

The secondary structure of DNA, which is the double helix, complexes with proteins to form a highly organized tertiary structure called chromatin. More than 50% of the mass of chromatin consists of proteins. About half of the protein mass of chromatin is made up of the positively charged proteins called histones (types—$H_1$, $H_{2A}$, $H_{2B}$, $H_3$, and $H_4$). Histones contain a high percentage (20–30%) of basic amino acids (lysine and arginine) that confer the positive charge and bind strongly to the negatively charged DNA. The other half of the protein mass of chromatin is made up of non-histone proteins, which are broadly classified as those having struc-tural roles (like chromosomal scaffold proteins) and functional roles (those involved in replication, transcription, regulation) (Pierce, 2005).

Chromatin has an extremely complex structure with several levels of the organi-zation (Fig. 10).

*LEVEL 1*—Isolated chromatin in a hypotonic solution appears as beads on a string when viewed under an electron microscope. Complete digestion of this chro-matin by micrococcal nuclease cleaves the string between the beads, chews up all the DNA in the string, and leaves individual beads or units called nucleosomes. Nucleosome, the simplest level of organization in chromatin, contains histone octamer or core (two copies each of $H_{2A}$, $H_{2B}$, $H_3$, and $H_4$) and 146 bp of DNA. The nucleosome core is a wedge-shaped disc around which DNA winds about 1.65 times in a left-handed direction resulting in negative supercoiling of DNA (Turner et al., 2005). The entry and exit of the DNA on the nucleosome core of histone octamer is sealed off by the binding of $H_1$ histone, which acts as a clamp and locks

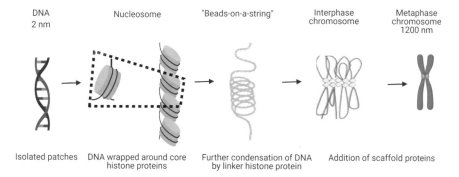

Fig. 10 Levels of organization of chromatin. Richard Wheeler at en.wikipedia, CC BY-SA 3.0 http://creativecommons.org/licenses/by-sa/3.0/, via Wikimedia Commons. (Source: Chromatin-structure-DNA-is-wrapped-around-a-histone-octamer-to-form-nucleosomes_W640.jpg)

the DNA in place. The nucleosome together with H1 and another 20–22 bp of DNA is called chromatosome, the next level of chromatin organization. Adjacent chromatosomes are separated by linker DNA that varies in size from 30 to 40 bp depending on the cell type (Annunziato, 2008). The non-histone proteins may bind to the linker DNA. This first level of chromosomal packing, which is 10 nm thick, is five to seven times more compact than free DNA.

*LEVEL 2*—The 30 nm fiber or nucleosome solenoid. Much of the chromatin that spills over by gently breaking open the nuclei is of 30 nm thickness. This thickness is achieved by winding the nucleosome chain into a higher-order left-handed helix or solenoid with six nucleosomes per turn. Conversely, chromatin fiber or a string of nucleosomes with intact $H_1$ are reported to come closer on increasing salt concentration and attain the 30-nm-thick solenoidal structure at 60 mM. This level of organization is obtained by the interaction of $H_1$ located in the central "hole" of the solenoid. Thus, H1 plays a fundamental role in forming the 30 nm fiber and achieving 40-fold packing (Robertis & Robertis, 2006).

*LEVEL 3*—A series of loops of the 30 nm fiber (each about 300 nm long, encompassing 10,000–20,000 bp of DNA) get anchored at their base to a protein scaffold or nuclear matrix, condensing the chromatin to 300 mm thickness. The 300 nm fiber is further folded to form a 250-nm-thick fiber which undergoes tight helical coiling to produce the 700-nm-thick metaphase chromosomes (Pierce, 2005).

## 3.5   Functions of DNA

1. DNA serves as the genetic material in most living organisms.
2. The small segments of DNA, known as genes, code for protein through an intermediary mRNA. Proteins (structural/functional) play an essential role in carrying out all the functions necessary for life, thus manifesting the genetic information in DNA.
3. Mutations or changes in the DNA sequence lead to new expressions and serve as the foundation for evolution.

## 4   RNA

In the year 1981, Thomas Cech and colleagues discovered that RNA can function as a biological catalyst. They found that one intron (400 nucleotides long) of *Tetrahymena thermophila* (protozoan) rRNA has the remarkable property to splice/excise itself in the absence of any proteins. They also found that the secondary structure of RNA is essential to drive this splicing. Over the years, many more examples of catalytic RNA performing various functions were discovered, and these RNA were called ribozymes. The discovery of ribozymes further strengthened the perception of RNA being the original genetic material. It is proposed that, around 3.5 million years ago, self-replicating ribozymes probably arose and led to the origin of life on earth (Pierce, 2005).

As life evolved in the RNA world, RNA served both as the genetic material and a catalyst. Over time, proteins that began to be synthesized on RNA took over the catalytic role, and the more stable DNA took over the genetic role, giving way to the DNA world. RNA still plays a vital role in the DNA transcription and protein synthesis of modern cells. RNA is primarily found in the cytoplasm, either free (like tRNA) or associated with ribosomes (rRNA, mRNA) and in the nucleolus (rRNA). RNA is also found in the nucleus (mRNA), mitochondria, and chloroplast.

## 4.1  Structure

Although there are multiple types of RNA, their basic structure is similar. RNA is a linear polymeric molecule formed by the stringing of the 5′P end of one ribonucleotide to the 3′OH end of the previous ribonucleotide by a phosphodiester bond (Fig. 11). Unlike DNA, RNA is usually single-stranded, having ribose sugar in its nucleotides and U instead of T as one of the pyrimidine bases. The single-stranded RNA does not follow Chargaff's rule; hence, purine to pyrimidine $\neq$ 1 or A/U $\neq$ G/C $\neq$ 1. RNA is folded on itself to form helices held together by

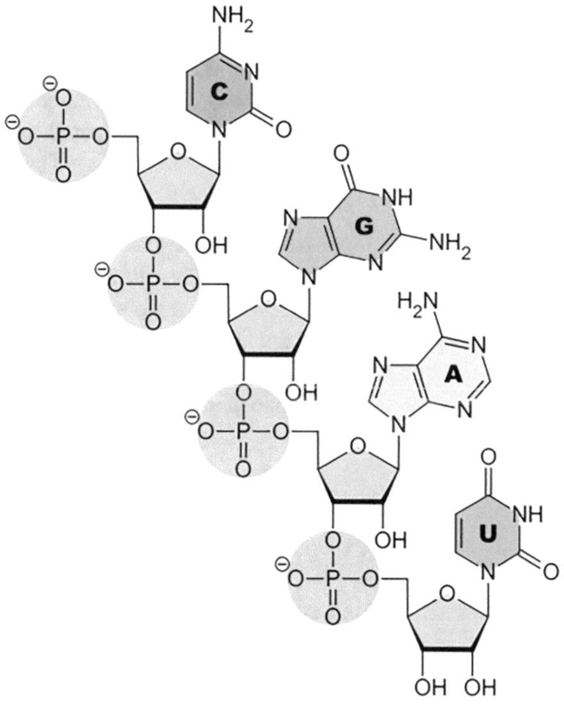

**Fig. 11** Primary structure of RNA. (Source: RNA-Nucleobases.png)

complementary base pairing forming stem-loop or hairpin-loop structures. This secondary structure provides stability to RNA and is critical for its function, like the binding of tRNA at the correct sequence of mRNA during translation (Robertis & Robertis, 2006).

## 4.2   Types of RNA

Depending upon the role RNA plays, it can be broadly categorized as genetic and non-genetic RNA.

1. *Genetic RNA*—When RNA acts as the genetic material in an organism, it is called genetic RNA. It is mainly found in viruses, particularly in plant viruses, and maybe single-stranded (ss) or double-stranded (ds). Examples of some RNA viruses are given in the following table.

|                      | Plant viruses | Animal viruses                          | Bacteriophages |
|----------------------|---------------|-----------------------------------------|----------------|
| Single-stranded (ss) | TMV           | Influenza, poliomyelitis, SARS-CoV-2    | MS2, F2, R17   |
| Double-stranded (ds) | Wound tumor   | Reovirus                                | –              |

2. *Non-genetic RNA*—In organisms where DNA is the genetic material, RNA plays other functional roles. Several kinds of RNA play an important role in gene expression. The various types and sizes of RNA present in the eukaryotic cell nucleus are collectively called heterogeneous nuclear RNA (hnRNA). The hnRNA includes pre-mRNA, mRNA (messenger RNA), rRNA (ribosomal RNA), tRNA (transfer RNA), snRNA (small nuclear RNA), miRNA (micro-RNA), and snoRNA (small nucleolar RNA). The significant chunk of hnRNA is constituted by pre-mRNA, which undergoes further processing to form mRNA.

– *mRNA* or *messenger RNA* was first discovered and named by Jacob and Monod in 1961. As the name rightly speaks, they carry the coding instruction from DNA to the ribosomes and serve as a template for protein synthesis. They carry the same base sequence as the DNA from which it is copied (template DNA), except that T is substituted by U. They constitute 5% of the total cellular RNA. They have a very short lifespan and wither away quickly, so their turnover rate is very high.

In prokaryotes, transcription is coupled to protein synthesis due to the lack of a nuclear membrane. This means that, as the mRNA is transcribed and its 5′ end is released from the RNA polymerase complex, it gets associated with ribosomes. The translation machinery is set on track for protein synthesis. Each amino acid of the protein is determined by a set of three nucleotides called codons. Hence, the 4-lettered base code of mRNA is translated into a sequence of 20 amino acids, forming a protein. In prokaryotes, all the different types of RNA are transcribed by a single RNA polymerase (Robertis & Robertis, 2006).

Whereas in eukaryotes, RNA polymerase II transcribes a large precursor RNA called pre-mRNA in the nucleoplasm. It is called pre-mRNA as it has to undergo extensive processing to form mRNA, which can be translated. As the pre-mRNA is transcribed, it gets associated with proteins forming ribonucleoproteins (RNPs) that confer stability. These proteins bind to specific sequences on the pre-mRNA and play an important role in posttranscriptional modifications to form mature mRNA (Turner et al., 2005). They also determine the nuclear export of mRNA and its association with ribosomes and contribute to translational regulation.

The formation of mRNA from pre-mRNA includes the following posttranslational modifications:

(a) 5′-capping—The 5′ end of pre-mRNA has three phosphates of which the terminal phosphate is cleaved, and 7-methylguanine forms a unique 5′-5′ triphosphate bridge during capping. Capping serves as a barrier to 5′ exonucleases and confers stability to the mRNA molecule. It also helps in the binding of mRNA to the ribosome and functions in the initiation of translation (Robertis & Robertis, 2006).
(b) 3′ cleavage—The pre-mRNA consists of sequences way beyond the coding region, and these are cleaved.
(c) Polyadenylation—Following the 3′ cleavage, 50–250 adenine nucleotides are attached by poly-A polymerase enzyme forming a poly-A tail. This tail promotes mRNA stability by resisting the 3′ exonuclease action.
(d) Splicing—Many eukaryotic genes contain coding regions called exons, and the noncoding regions are called intervening sequences or introns. A pre-mRNA is constituted of both introns and exons. To obtain a functional mRNA, introns must be precisely excised and the molecule relegated, thus joining exons. This is called splicing, and it occurs within a large complex called the spliceosome. The spliceosome is composed of small nuclear ribonucleoprotein particles (snRNPs, formed by the association of snRNAs with proteins) and pre-mRNA (Pierce, 2005).

Based on the number of cistrons (the region of mRNA that codes for a single protein), mRNA can be monocistronic (all eukaryotic mRNA) or polycistronic (prokaryotic mRNA code for more than one protein) (Fig. 12).

– *rRNA or ribosomal RNA* constitutes about 80% of total cellular RNA. It forms a major component of ribosomes and plays a catalytic and structural role in protein synthesis (translation). The rRNA exhibits a high degree (about 70%) of secondary structure, which helps in maintaining the ribosome structure. The rRNA provides a 3-D matrix for the sequential binding of enzymes involved in translation. It is also actively involved in recognizing the conserved sequences of mRNA and tRNA during translation (Pierce, 2005).

Prokaryotes have 23S, 16S, and 5S rRNA in their ribosomes, which are named according to their rate of sedimentation measured in Svedberg units "S." These rRNAs are encoded by a single rDNA by RNA polymerase I. Mature rRNA is

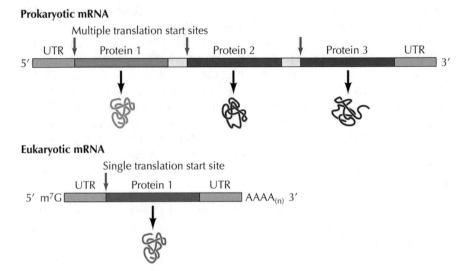

**Fig. 12** Polycistronic prokaryotic mRNA and monocistronic eukaryotic mRNA. (Source: main-qimg-d0026c2aa9c807fdcd07d050508abc3d-c)

**Fig. 13** Processing of pre-rRNA primary transcripts in eukaryotes to form mature rRNA

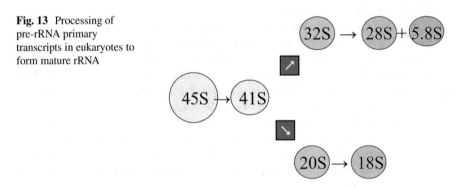

formed by the cleavage and modification of initial transcripts, and soon after, they associate with proteins to form ribosomes. Prokaryotic ribosomes (the 70S) are made of two subunits, the larger 50S (containing 23S and 5S rRNA) and the smaller 30S (containing 16S rRNA). Ribosomes provide a scaffold for the ordered interaction of molecules during translation (Pierce, 2005).

Eukaryotic ribosomes (80S) are made of two subunits, the larger 60S (containing 28S, 5.8S, and 5S rRNA) and the smaller 40S (containing 18S rRNA). The 28S, 5.8 S, and 18S rRNA are transcribed as a single 45S ribosomal precursor RNA (pre-rRNA) in the nucleolus by RNA polymerase I. The 45S pre-rRNA folds and complexes with proteins as it is transcribed in the nucleolus. These proteins, known as small nucleolar ribonucleoproteins (snRNPs), methylate the pre-rRNA on the 2′-ribose moiety of 28S and 18S rRNA. The non-methylated regions of the 45S pre-rRNA are cleaved, forming smaller components, as shown in Fig. 13 (Robertis & Robertis, 2006).

Whereas 5S rRNA is transcribed outside the nucleolus by RNA polymerase III, it is transported to the nucleolus to be incorporated into the ribosome. The nucleolus is an aggregation of rDNA looped out of the nucleolar organizer region of chromosomes, transcribed rRNA, and proteins. It is here in the nucleolus that the early assembly of ribosomes occurs. To fulfill the need for a large number of rRNA, chromosomes contain multiple copies of rRNA coding genes in tandem repeat, separated by spacer DNA (Robertis & Robertis, 2006).

– *t-RNA or transfer RNA* is a family of about 60 small-sized RNA molecules which recognize mRNA codons and carry activated amino acids to the ribosomes for protein synthesis or translation. They are also called adapters or supernatant or soluble RNA and constitute about 15% of the total cellular RNA. They are the smallest among the major RNA types, about 75–85 nucleotides long, and have a sedimentation rate of 4S. tRNA is transcribed by RNA polymerase III (Robertis & Robertis, 2006). The RNA is folded into a cloverleaf-shaped secondary structure resulting from the formation of about four stem-loop structures held together by complementary base pairing (Fig. 14).

A 7 bp double-helical stem is formed near the 5′ and 3′ ends. The 5′ end has a free phosphate group, and the 3′ end (amino acid acceptor end) overhangs as

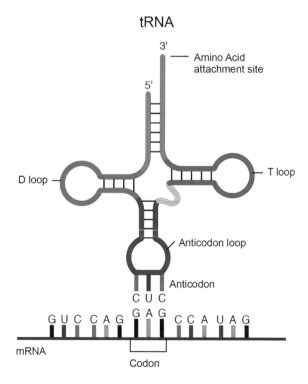

**Fig. 14** Structure of tRNA. (Source: qZgsfkZYSei9aS8kPsFS_tRNA.jpg)

5'CCA3', where a particular amino acid is attached covalently by aminoacyl-tRNA synthetase. The amino acid to be carried is determined by the anticodon sequence (base triplet complementary to the codon) present in the anticodon loop. The anticodon and codon pairing bring the correct amino acid to the precise position in the translation machinery, where it gets stitched to the growing end of the protein/polypeptide chain.

The D arm has a D loop (DHU or dihydrouracil loop) containing the modified base and a 3–4 bp stem. The D loop plays an important role in stabilizing the tertiary structure of tRNA. The T arm is composed of a 5 bp stem and a T loop (TΨC loop; Ψ denotes pseudouridine). The T loop has modified bases and a conserved sequence (TΨCG) which helps in the interaction of tRNA with the ribosome (Robertis & Robertis, 2006).

– Other Small RNA.

Eukaryotic cells also contain small, stable RNA like any other ribonucleotide chain but do not code for any protein. These RNAs instead associate with proteins and form ribonucleoprotein (RNP) particles consisting of eight small, basic core proteins and one RNA molecule. Depending on their localization within the cell, they may be small nuclear RNP (*snRNPs*) or small cytoplasmic RNP (*scRNPs*), and small nucleolar ribonucleoproteins (*snoRNPs*). The snRNPs, as described previously, bind to the pre-mRNA and play an important role in posttranscriptional modifications, whereas snoRNPs bind to pre-rRNA and play an essential role in its processing (Turner et al., 2005). Not much information is available about *scRNPs*.

*miRNA* or microRNAs are about 22 nucleotides long and play a role in eukaryotic gene regulation. They inhibit or silence gene expression by binding to mRNA and inhibit translation or protein synthesis. They play a significant role in cancer and other diseases. 25% of the human miRNA genes are located in the introns of protein-coding genes, while others appear to be transcribed as a polycistronic unit (Turner et al., 2005).

*Ribozymes* are catalytic RNA that may be sometimes associated with auxiliary proteins. They are found in many organisms across all the kingdoms of life. They catalyze biological reactions like RNA splicing, RNA cleavage, and protein synthesis. In the last two decades, many ribozymes have been developed in vitro also.

## 4.3   Functions of RNA

It can be understood from the discussion on RNA that they play multiple roles in the functioning of cells. These functions can be summarized as follows:

1. In some plant and animal viruses, they serve as the genetic material.
2. mRNA, which is the coding RNA, serves as a template for protein synthesis.
3. rRNAs along with the cluster of proteins form ribosomes, the seat of protein synthesis.

4. tRNA serves as an adaptor molecule by bringing the amino acids on the mRNA, as amino acids cannot establish bonds with the codons of mRNA.
5. The RNPs bind to RNA transcripts and bring about their posttranscriptional modifications.
6. miRNA and siRNA silence gene expression.
7. Ribozymes catalyze reactions all by themselves.
8. snoRNA binds to the pre-rRNA and designs the formation of the 28S, 5.8S, and 18S rRNA.

# References

Annunziato, A. (2008). DNA packaging: Nucleosomes and chromatin. *Nature Education, 1*(1), 26. https://www.nature.com

Berg, J. M., Tymoczko, J. L., Stryer, L., & Stryer, L. (2002). *Biochemistry* (5th ed.). W.H. Freeman.

Griswold, A. (2008). Genome packaging in prokaryotes: The circular chromosome of E. coli. *Nature Education, 1*(1), 57. https://www.nature.com

Ojaghi, A., Carrazana, G., Caruso, C., Abbas, A., Myers, D. R., Lam, W. A., & Robles, F. E. (2020). Label-free hematology analysis using deep-ultraviolet microscopy. *PNAS, 117*(26), 14779–14789. https://doi.org/10.1073/pnas.2001404117

Pierce, B. A. (2005). *Genetics: A conceptual approach* (2nd ed.). W.H. Freeman.

Pray, L. A. (2008). Discovery of DNA structure and function: Watson and crick. *Nature Education, 1*(1), 100. https://www.nature.com

Robertis, E. D. P. D., & Robertis, E. M. F. D. (2006). *Cell and molecular biology* (8th ed.). Lippincott William & Wilkins.

Turner, P., McLennan, A., Bates, A., & White, M. (2005). *Molecular Biology* (3rd ed.). Taylor & Francis.

Ussery, D. W. (2002). DNA structure: A-, B- and Z-DNA Helix Families. In *eLS*, (Ed.). https://doi.org/10.1038/npg.els.0003122.

Watson, J. D., Baker, T. A., Bell, S. P., Gann, A., Levine, M., & Losick, R. M. (2004). *Molecular biology of the gene* (5th ed.). Pearson Education.

# General Steps During Isolation of DNA and RNA

**Abstract** The genetic material in any organism passes from one generation to the next generation through the different ways of reproduction. This hereditary material is present either as DNA or RNA inside cells. The biological study of these DNA or RNA molecules for various purposes requires their isolation from their native location to the in vitro conditions. Over the years, scientists have developed different extraction techniques for DNA and RNA from a variety of cells and organisms. Moreover, with the advancement of technologies, these techniques have undergone tremendous sophistication and automation for the quick and economical yield of DNA and RNA. In spite of different methodologies used to isolate nucleic acids from the cells, there are few common steps and general biochemicals which are regularly used in the isolation steps. This chapter provides an overview of all those general steps along with the basic principles behind applying these steps during isolation protocols.

**Keywords** DNA · RNA · Nucleic acid · Isolation steps · Extraction

## 1 Introduction

Nucleic acids (DNA or RNA) are the hereditary molecules that harbor biological instructions for making each species unique. This genetic material is passed on from one generation to another at the time of reproduction. Each organism contains multiple molecules of DNA per cell. In eukaryotes, DNA is found localized in a particular region of the cell known as the nucleus. This DNA is packaged tightly in the form of chromosomes. The complete set of nuclear DNA of an organism is called a genome.

Extraction or isolation of DNA is an initial step for various molecular biology, genetics, and recombinant DNA technology applications. It is one of the basic steps to study a genome or identify gene sequences in a heterogeneous gene pool. The isolated and purified nucleic acids are the starting point for accurate and practical

© The Author(s), under exclusive license to Springer Nature Switzerland AG 2022
A. Gautam, *DNA and RNA Isolation Techniques for Non-Experts*, Techniques in
Life Science and Biomedicine for the Non-Expert,
https://doi.org/10.1007/978-3-030-94230-4_2

exploration of downstream processes in a cell. Moreover, DNA isolation is of prime importance in medical biology to study the genetic cause of a particular disease and develop its diagnostics, vaccines, and drugs. It is also vital in various diversified fields, including forensic sciences (for crime investigations), determining paternity, genetic engineering for modifications in plants and animals, DNA-based cloning, and blotting. The isolation protocols have gained tremendous importance with the advent of various high-throughput genomic techniques such as genomic sequencing, construction of gene libraries, DNA fingerprinting, restriction fragment length polymorphism (RFLP), etc. All the areas mentioned above of biology usually require DNA in its purest form because the quality and integrity of the nucleic acid directly affect the results of all succeeding scientific experiments (Tan & Yiap, 2009).

Similarly, RNA extraction from a cell or tissue is termed "RNA isolation." As RNA is the photocopy of DNA in almost all organisms, the isolated RNA is used to study and manipulate the various cellular processes. For example, CRISPR-Cas and RNAi (RNA interference) technologies are widely used for regulating the expression of genes in an organism. Moreover, the whole study of "reverse genetics" is based on the applications of isolated RNA. During the present COVID-19 crisis, RNA isolation is a crucial pre-analytical step for detecting the level of SARS-CoV-2 through RT-PCR (reverse transcription polymerase chain reaction). Therefore, different robust, simple, and cost-effective methods are required for the large-scale study of DNA and RNA in different labs and fields.

## 2    Common Steps of Nucleic Acid Isolation

Several methods can do DNA and RNA isolation, but choosing the best one can save time and energy. They can be isolated from biological material such as living or conserved tissues, cells, or viruses. The following factors need to be considered before choosing the suitable method:

(a) **Starting sample**: Different techniques of DNA and RNA extraction are used depending on the used sample (e.g., bacteria, virus, human tissues, soil, leaf extract, forensic samples, etc.).

(b) **Planned usage**: All biological applications (like PCR, sequencing, fingerprinting, SDS-PAGE, etc.) require different qualities or quantities of nucleic acid, and therefore, one should choose the isolation protocols as per their planned use.

(c) **Sample quantity**: The method selection will also depend on the size/volume of the sample being analyzed. The abundance and cost of a particular sample are the two critical deciding factors for selecting a particular type of isolation protocol.

Primarily, effective nucleic acid purification requires four key steps (Shen, 2019):

1. Disruption of cells and tissues: The sample is first subjected to cell lysis to remove different lipid membranes around the nucleic acids. This helps in the

isolation of DNA or RNA into an aqueous solution. General procedures for cell lysis include enzymatic (by incubations with enzymes that hydrolyze cell membrane and cell wall components), thermal, chemical (by different detergents to solubilize lipid membrane components or chaotropic agents which perforate cell membranes by denaturing transmembrane proteins of the cells), or mechanical treatment (by freeze-thawing, grinding, or bead-beating), or an amalgamation of those (Lever et al., 2015). Additionally, chemical carriers (inorganic phosphate species) are put in various procedures to cover charged surfaces preceding the cell disruption, as both DNA and RNA adsorb to positively charged surfaces.

2. Denaturation of nucleoprotein complexes: After the first step, the sample is treated with protease to denature the nucleoprotein complexes and inactivate the nucleases (like RNase during RNA isolation).

3. Removal of contaminants: All other cellular contaminants like RNA during DNA isolation and DNA during RNA isolation are removed by RNase and DNase treatment, respectively. Further, the high quality and high purity of the target nucleic acid are maintained by removing protein, carbohydrate, lipids, etc. There are several purification steps (including precipitation) for the elimination of organic and inorganic molecules from subsequent aqueous extracts. These steps are followed by washing (with organic solutions such as phenol, chloroform, or cetyltrimethylammonium bromide) or column-based purification to eliminate extra cell components, degrading enzymes, or inhibitory substances.

4. Precipitation of the nucleic acid: The final step involves the precipitation of DNA or RNA molecules from the reaction mixture. This helps in the recovery of nucleic acids from the solution. Generally, precipitation with polyethylene glycol, isopropanol, and ethanol and filtration through ion-exchange resins, magnetic beads, silica columns, or gels are preferred methods (Lever et al., 2015).

The difference between the stability of DNA and RNA accounts for different extraction methods. The presence of the 2'-OH group on the pentose ring of RNA makes it very reactive, and hence it is more susceptible to hydrolysis. Also, RNA has the more reactive nitrogenous base uracil compared to thymine in DNA.

The extraction process has remained unchanged for decades; however, the modifications in chemistries and coatings have always been central to the research community. Scientific literature is flooded with details of the different DNA/RNA extraction methods that are currently in use, right from the modern-day spin-column kits to the old-fashioned methods like the paper strip method (Shendure et al., 2017). However, the amount and quality of the DNA or RNA obtained by each method are variable (Oliveira et al., 2014). Various procedures for membrane disintegration and isolation of DNA or RNA were discovered in characterization studies that evolved during the twentieth century and serve as the understructure for currently being practiced (Fig. 1).

However, the employment of different techniques is limited due to various disadvantages, including toxic chemicals such as phenol or chloroform, low yields, high cost, complexity, the inclusion of used chemicals as contaminants in extracted samples, and low throughput. Though methods are well-approved and generate

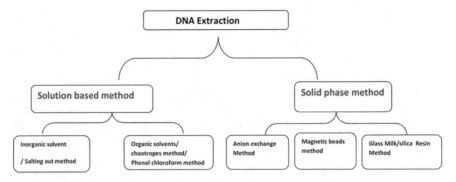

**Fig. 1** Different types of DNA isolation methods

high-quality nucleic acid, they are repeatedly painstaking, arduous, and too expensive or suffer from inadequate and erratic quantities of nucleic acids. These drawbacks are yet more substantial when considering applications of molecular diagnostics in low-income countries. Therefore, an option between the availability of advanced laboratory facilities with the researcher and the feasible and adaptable isolation methods must be selected before starting the DNA or RNA isolation (Martzy et al., 2019).

There are two important considerations that one should remember before starting the extraction of DNA or RNA: (1) the consumables available in the lab and (2) the yield/purity of nucleic acids required after extraction (Sirakov, 2016). The second consideration is of more importance and much scientific interest. To achieve this, agarose gel electrophoresis followed by a DNA or RNA extraction method is most used. This determines the relative quality and quantity of the recovered DNA or RNA (Mićić, 2016). Additionally, there are organic extraction, solid-phase extraction, Chelex extraction, and solution-based extraction techniques for DNA isolation, including centrifugation, silica-based extraction, magnetic bead-based methods contingent on the field of application, and the milieu in which the cells are examined.

## 3   Widely Used Extraction Methods

1. **Organic extraction**: Phenol-chloroform-isoamyl (PCI) alcohol extraction method is widely used in isolating nucleic acid. It is based on the liquid-liquid extraction and solubility of biomolecules. The following steps are applied during such extraction:

   (a) The first step is the cell lysis. The well-organized cells are disrupted, and cellular fragments are typically discarded after centrifugation. Phenol is a nonpolar solvent; DNA being polar dissolves in polar solutions and can be extracted using phenol.

(b) Proteins, lipids, carbohydrates, and cell debris are removed through the extraction of the aqueous phase with the organic mixture of phenol and chloroform (Tan & Yiap, 2009). Other types of nucleic acid, i.e., RNAs, are removed by treatment of the RNase enzyme.

(c) The hydrophobic layer would be settled at the bottom and the hydrophilic layer on top after centrifugation.

(d) DNA can then be precipitated from the supernatant by adding ethanol or isopropanol in the presence of a high concentration of salt.

(e) DNA precipitate is collected by centrifugation, and excess salt is rinsed with 70% ethanol and centrifuged to discard the ethanol supernatant.

(f) The DNA pellet is then dissolved with TE buffer or sterile distilled water.

2. **Silica-based technique**: Silica membranes/beads/particles are being used in combination with certain salts to which DNA selectively adsorbs. The contaminants are removed in wash steps, and chaotropic salts are used for protein denaturation.

3. **Magnetic separation**: This method uses DNA or RNA binding antibodies firmly attached to the magnetic beads or solid surfaces. After DNA binding, these beads are detached from other impurities, washed, and eluted by following the ethanol extraction protocols. This technique is fast and robotic but more expensive than other procedures of isolation.

4. **Anion exchange technology**: This method is based on the specific interaction between negatively charged phosphates of the DNA and positively charged surface molecules on the substrate. This method is specifically employed in plasmid DNA purification.

# 4 DNA Extraction and Purification Based on Source Materials

## 4.1 DNA Isolation from Microbes

For DNA isolation from bacteria, healthy cells are first allowed to grow in suitable broth media until they reach an exponential growth phase and then gathered. These cells are then ruptured using chemical reagents such as lysozyme, proteinase K, EDTA, and other detergents. Organic extraction or silica-based technologies are applied next to separate the DNA and protein components. Finally, the DNA is precipitated at a high concentration (Dhaliwal, 2013).

## 4.2 DNA Extraction from Animal Cells and Tissues

The basic steps are the same as in microbes; however, cell culture and preparation are generally diverse for the animal cells and tissues, but they are easier to lyse (Dhaliwal, 2013). Common lysing agents here are detergents. The initial step often needs to be mechanically homogenized as they might be intact. PCI method and ethanol precipitation could be employed for DNA isolation from animal tissues.

## 4.3 DNA Extraction from Plant Tissues and Cells

Due to cell walls, the same chemicals/lysozymes as used in previous cases are not appropriate for the lysis of the plant cells. Due to more polyphenols and polysaccharides in plant cells, simple phenol extraction cannot be used for DNA extraction. Therefore, a detergent known as CTAB (cetyltrimethylammonium bromide) is used, which forms an insoluble complex with nucleic acid and selectively precipitates DNA, leaving behind carbohydrates, proteins, and other contaminating components. Thereafter, the DNA-containing precipitate can be decomplex by dissolving it in NaCl (Dhaliwal, 2013). The alternative technique is to employ guanidinium thiocyanate (GITC), which helps in the denaturation and dissolving of proteins. GITC also dissociates nucleoproteins from the DNA and can be used in any tissue. In the presence of GITC, DNA also strongly binds to silica particles. Thus, the cell extract or tissue homogenate mixed with GITC can be applied to the chromatography columns packed with silica for DNA isolation. DNA then selectively binds to the column and can be easily eluted.

## 5 RNA Isolation

Compared to DNA isolation, RNA isolation is a little bit tricky because RNA extraction requires good laboratory practices for an RNase-free environment, and RNA has a very short half-life once extracted from the tissues. Three types of RNA occur naturally, which include ribosomal RNA (rRNA), messenger RNA (mRNA), and transfer RNA (tRNA). It is particularly very important to safeguard the RNA against degradation; otherwise, the transcripts with a very low copy number could be lost and will not be detected in the analysis steps. As stated earlier, RNA is especially unstable due to the pervasive presence of RNases which are enzymes present in the blood, all tissues, as well as most bacteria and fungi in the environment. Degraded RNA molecules fail to bind to the complementary sequences effectively. So, it is very much recommended to use fresh tissue samples.

To isolate intact RNA, strong denaturants are always used to inhibit endogenous RNases (Aderiye & Oluwole, 2014). RNase enzyme is heat-stable and refolds following heat denaturation. As RNase does not require cofactors, they are tough to inactivate. Therefore, in RNA extraction procedures, it is mandatory to ensure an RNase-free fraction, and a means to cool down the sample quickly. To effectively isolate RNA, a strong protein denaturant is required as it would degrade the RNase. A simple RNA extraction protocol is done using guanidinium thiocyanate-phenol-chloroform. Different methods used to isolate RNA might lead to different results, and therefore a robust RNA isolation method is required for reproducibility. Researchers should optimize these methods for their specific application and keep in mind that "total RNA" extraction methods do not isolate all types of RNA equally (Brown et al., 2018).

Basic steps in RNA extraction are the following:

1. The samples are first homogenized and lysed in guanidinium thiocyanate solution.
2. To this phenol, sodium acetate and chloroform are added, and the mixture is kept for centrifugation.
3. The mixture separates into an aqueous layer, interphase, and an organic layer, with the RNA in the aqueous layer.
4. The top layer is extracted, and the RNA is precipitated using an isopropanol solution.

# References

Aderiye, B. I., & Oluwole, O. A. (2014). How to obtain the organelles of prokaryotic and microbial eukaryotic cells. *Journal of Cell and Animal Biology, 8*(6), 95–109.

Brown, R. A., Epis, M. R., Horsham, J. L., Kabir, T. D., Richardson, K. L., & Leedman, P. J. (2018). Total RNA extraction from tissues for microRNA and target gene expression analysis: Not all kits are created equal. *BMC Biotechnology, 18*(1), 1–11.

Dhaliwal, A. (2013). DNA extraction and purification. *Mater. Methods, 3*, 191.

Lever, M. A., Torti, A., Eickenbusch, P., Michaud, A. B., Šantl-Temkiv, T., & Jørgensen, B. B. (2015). A modular method for the extraction of DNA and RNA, and the separation of DNA pools from diverse environmental sample types. *Frontiers in Microbiology, 6*, 476.

Martzy, R., Bica-Schröder, K., Pálvölgyi, Á. M., Kolm, C., Jakwerth, S., Kirschner, A. K., … Reischer, G. H. (2019). Simple lysis of bacterial cells for DNA-based diagnostics using hydrophilic ionic liquids. *Scientific Reports, 9*(1), 1–10.

Mićić, M. (Ed.). (2016). *Sample preparation techniques for soil, plant, and animal samples.* Humana Press.

Oliveira, C. F. D., Paim, T. G. D. S., Reiter, K. C., Rieger, A., & D'azevedo, P. A. (2014). Evaluation of four different DNA extraction methods in coagulase-negative staphylococci clinical isolates. *Revista do Instituto de Medicina Tropical de São Paulo, 56*, 29–33.

Shen, C. (2019). Extraction and purification of nucleic acids and proteins. In C.-H. Shen (Ed.), *Diagnostic Molecular Biology* (pp. 143–166). Academic Press.

Sirakov, I. N. (2016). Nucleic acid isolation and downstream applications. *Nucleic Acids-From Basic Aspects to Laboratory Tools*, 1–26.

Shendure, J., Balasubramanian, S., Church, G. M., Gilbert, W., Rogers, J., Schloss, J. A., & Waterston, R. H. (2017). DNA sequencing at 40: Past, present and future. *Nature, 550*(7676), 345–353.

Tan, S. C., & Yiap, B. C. (2009). DNA, RNA, and protein extraction: The past and the present. *Journal of Biomedicine and Biotechnology, 2009.*

# Part II
# Methods

# Phenol-Chloroform DNA Isolation Method

**Abstract**  Isolation of DNA through phenol-chloroform is one of the most commonly used methods in molecular biology. Researchers can obtain a good yield of high molecular weight DNA from any biological sample using a mixture of phenol-chloroform-isoamyl alcohol. The basic principle involves the separation of DNA, RNA, and proteins in a sample, based on the differential solubilities of these molecules in different immiscible liquids. The isolated DNA can be used in gene amplification, fingerprinting, single nucleotide polymorphism studies, RFLP, and gene quantification. Though the present chapter describes the isolation of DNA from blood cells, the basic steps are similar for DNA isolation from other tissues or cells, with few minor variations in the initial steps. Despite all these advantages, this method uses different dangerous organic solvents and hence should be carefully handled. Moreover, this method is laborious and time-consuming compared to the other methods available, for example, column-based isolation of DNA.

**Keywords**  DNA · Phenol · Chloroform · Protein · RNase · Genetic engineering · Isopropanol · Ethanol

## 1  Introduction of the Technique

For obtaining good quality and original data, it is essential to obtain pure DNA as the quality, quantity, and integrity of the DNA will directly affect the results of the experiment and the quality of the data. The phenol-chloroform DNA isolation method is one of the most classical and widely used methods to obtain a high molecular weight DNA such as human genomic DNA. This method is widely used in single nucleotide polymorphism (SNP) studies, gene quantification, gene amplification, DNA fingerprinting, and RFLP. DNA from blood cells, soft tissues like muscle, brain, flower, hair follicles, dental pulp tissue, epithelial cells, and osteocytes of bones can be isolated easily through this method.

© The Author(s), under exclusive license to Springer Nature Switzerland AG 2022    33
A. Gautam, *DNA and RNA Isolation Techniques for Non-Experts*, Techniques in
Life Science and Biomedicine for the Non-Expert,
https://doi.org/10.1007/978-3-030-94230-4_3

# 2    Basic Principle

The phenol-chloroform DNA isolation method is also referred to as organic extraction or liquid-liquid extraction. The principle of this method utilizes the separation of DNA, RNA, and protein-based on differential solubilities of these molecules in different immiscible liquids. It involves using sodium dodecyl sulfate (SDS) to disrupt the cell membrane and proteinase K to enzymatically digest and remove proteins bound to DNA to pack them into chromosomes, respectively. Subsequently, an equal volume of phenol-chloroform solution is added to the lysed cells, forming two phases after centrifugation: the upper aqueous and lower organic phases. This allows the separation of nucleic acids from proteins, the former being more soluble in the aqueous phase (Fig. 1). It is essential to keep in mind that the pH of the phenol must be kept slightly alkaline (pH 8) because the alkaline pH keeps the nucleic acids (DNA and RNA) negatively charged, thus allowing the portioning of both into the aqueous phase. At acidic pH, the negative charge on the DNA, imparted by the phosphate diesters in nucleic acids, is neutralized by protonation. On the contrary, RNA escapes neutralization in acidic pH and remains in the aqueous phase as single-stranded. It contains exposed nitrogenous bases that can form hydrogen bonds with water (Brawerman et al., 1972). Thus, the pH of phenol must be kept slightly alkaline to inhibit its entrapment at the aqueous-organic interphase (Bradley et al., 2001).

# 3    Protocol

## 3.1    Reagents Required

- **EDTA 0.5 M solution, pH 8.0** (to be stored at room temperature)
- **Sodium chloride, 5 M solution** (to be stored at room temperature)
- **Lauryl sulfate, 10% solution** (10% SDS, 10% sodium dodecyl sulfate) (to be stored at room temperature)
- **Tris-EDTA Buffer 100X Concentrate** (1 M Tris-HCl, 0.1 M EDTA, pH around 8.0) (to be stored at room temperature)
- **Ribonuclease A**, free of DNase activity (to be stored at 2–8 °C)
- **Sodium acetate, 3 M solution, pH 5.2** (to be stored at room temperature)

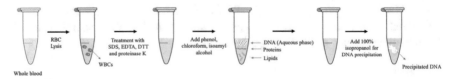

**Fig. 1** Outline of phenol-chloroform extraction method for DNA extraction from the whole blood sample

- **Phenol-chloroform-isoamyl alcohol 25:24:1, saturated with 10 mM Tris, pH 8.0, 1 mM EDTA**
- **Isopropanol (isopropyl alcohol), 100%**
- **Ethanol (ethyl alcohol), 95%**

## 3.2   Working Solutions

| Solution | Preparation | Storage |
|---|---|---|
| RBC Lysis Solution (1 mM EDTA), 1000 mL | Dilute 2 mL of EDTA 0.5 M solution in 998 mL deionized water, pH 8.0 | Room temperature |
| NaCl solution, 1 M, 1000 mL | Dilute 200 mL of sodium chloride, 5 M solution in 800 mL deionized water | Room temperature |
| Cell Lysis Solution (10 mM Tris-HCl, 26 mM EDTA, 17.3 mM (0.5%) SDS), 1000 mL | Dilute 10 mL Tris-EDTA Buffer 100X Concentrate and 50 mL EDTA 0.5 M solution and 50 mL lauryl sulfate, 10% solution in 890 mL deionized water, pH 7.3 | Room temperature |
| RNase A Solution (4 mg/mL) | Dissolve 10 mg RNAase in 2.5 mL TE buffer | 2–8 °C |
| Sodium acetate solution, 3 M | Use as purchased, no preparation needed | 2–8 °C |
| Phenol-chloroform-isoamyl alcohol 25:24:1, saturated with 10 mM Tris, pH 8.0, 1 mM EDTA | Use as purchased, no preparation needed | 2–8 °C |
| Isopropanol (isopropyl alcohol), 100% | Use as purchased, no preparation needed | Room temperature |
| Ethanol (ethyl alcohol), 70% Mix well | Add 737 mL of 95% ethanol (ethyl alcohol), and then add deionized water to bring the total volume to 1000 mL | 2–8 °C |

## 3.3   Procedure

### 3.3.1   Sample Collection

The sample (tissue/cells) must be properly collected and stored for DNA extraction. Freshly collected samples are recommended for obtaining good-quality DNA. The tissues must be isolated and minced quickly, followed by freezing in liquid nitrogen to minimize the activity of endogenous nucleases. Moreover, the tissues must be properly homogenized and well dispersed to allow quick and efficient access to proteinase K and SDS. A standard protocol for the isolation of DNA from the blood cells has been mentioned below, though the procedure is similar for all other samples.

For isolating DNA from whole blood samples, the blood must be collected in vials with anticoagulants (EDTA vials). Heparin should not be used for blood collection as it hinders polymerase chain reaction (Beutler et al., 1990). Freshly collected blood samples are recommended for DNA isolation; however, blood must be stored at 0 or −70 °C before DNA isolation if required.

### 3.3.2   Protocol for DNA Isolation from Whole Blood (20 mL)

1. Mix the 20 mL blood sample well by inversion.
2. Pour the blood sample in a 50 mL canonical centrifuge tube and add 25 mL of RBC Lysis Solution, close the tube, and mix the solution well by invert mixing (5 times); incubate the solution for 2 min at room temperature, and mix again by inversion (5 times). **Note**: *Rinse the transfer pipette and tube containing blood sample with the RBC Lysis Solution to maximize yield.*
3. Centrifuge the tube at 2000×*g* for 10 min at room temperature.
4. Take off the supernatant carefully, leaving behind the white pellet containing WBCs along with a bit of residual liquid. **Note**: *Do not touch or disturb the white pellet with micro tips. Discard the supernatant in a biohazard container.*
5. Resuspend the pellet containing WBCs into the residual liquid by slight tapping till the pellet is resuspended.
6. Add 20 mL of RBC Lysis Solution, mix by invert mixing, and centrifuge again at 2000×*g*. Take off the supernatant carefully, leaving the pellet behind, and discard in a biohazard container.
7. Add 1.0 mL NaCl solution, 1 M, and mix by pulse vortexing to resuspend WBCs. If there are any clumps, vortex briefly till the clump is dispersed. **Note**: *Hash and continuous vortexing can damage WBCs.*
8. Add 6.0 mL of Cell Lysis Solution to resuspended WBCs.
9. Add 50 μL RNase A solution to the tube and incubate at 37 °C for 60 min to allow complete lysis of the WBCs and RNA degradation. **Note**: *Samples are stable in Cell Lysis Solution for at least 18 months at room temperature. Therefore, at this step, the DNA extraction process can be stopped for few days if required.*
10. Add 1.0 mL sodium acetate solution, 3 M to the tube followed by brief vortexing.
11. Add 1.0 mL of phenol-chloroform-isoamyl alcohol 25:24:1—vortex by pulse vortexing for around 1 min. Mix the solution by inversion.
12. Centrifuge the tube at 2000×*g* for 10 min at room temperature to separate phases.
13. Allow the tubes to stand for 15 min or until the upper layer turns clear. Transfer the upper phase, using a cut tip cut-tips (tips cut around 2 mm from the rear end) to a 15 mL conical centrifuge tube containing 7.0 mL of isopropanol, 100%. Avoid touching the other phases with a pipette. The phases should not mix during aspiration of the aqueous phase.
14. Close the 15 mL tube tightly and invert it 20–30 times slowly to facilitate precipitation of the DNA. If visible DNA precipitates (it will appear like a white clump), you may proceed directly to the next step. However, if no precipitate is seen, leave the samples to stand overnight at 4 °C.
15. Spin at 2000×*g* for 5 min at room temperature.

16. Take off the supernatant completely, leaving the intact DNA pellet in the tube. Add 10 mL of ethanol, 70% to the sample. Invert numerous times to wash the DNA pellet.
17. Spin at 2000×$g$ for 3 min at room temperature.
18. Remove the ethanol carefully with the DNA pellet intact. Invert the tube and keep it on a blotting sheet. Make sure the pellet is not dislodged. Let the pellet air-dry for around 20 min or until no moisture drops are visible in the tube.
19. Add 1 mL of TE buffer to the tube. (If the starting sample volume was less than 10 mL, add only 500 μL TE buffer.) Close the cap tightly. Vortex for 5–10 s. Keep the sample in an incubator at 37 °C overnight to dissolve the DNA pellet completely. Slightly tap to mix.
20. When the DNA pellet is dissolved completely, the DNA sample is ready for quantitative and qualitative analysis, storage, and experimentation.

# 4   Precautions

• It is essential to wear a lab coat, safety glasses, and gloves while handling phenol as it is being severely acidic; it may cause severe burns in the skin and damage clothes. Furthermore, it is recommended to redistill phenol before use as the oxidation products of phenol may break the nucleic acid chain. All the processes which require phenol should be executed in the fume hood. In case of any spill-age over the skin or clothes, the skin must be washed immediately and thor-oughly with water. If needed, emergency medical assistance must be sought in case of extensive burns.
• The laboratory area must be thoroughly cleaned with 70% alcohol throughout all the procedures of the experiment.
• As ethanol is highly combustible, it should be handled very carefully.
• The tubes must be properly balanced during centrifugation.
• While aspirating aqueous phase containing DNA, wide-bore cut-tips (tips cut around 2 mm from the rear end) should be used to avoid mechanical disruption and shearing of DNA during pipetting.
• While aspirating the upper aqueous phase from the lower organic phase, it must be ensured that the two phases do not get mixed at the time of pipetting. In case the two-phase gets mixed, it should be centrifuged again for the separation of two phases.
• Filter tips should be used for pipetting to avoid cross-contamination.

# 5   Applications

## 5.1   Gene Sequencing

The most common application of DNA extraction is sequencing a genome, either for identifying or classifying a newly found species or studying genetic variations in a living organism. Many human diseases are associated with genetic mutations and variations in single nucleotide polymorphisms (SNPs). Some diseases diagnosed by gene sequencing include Down's syndrome, Huntington's chorea, sickle-cell anemia, Duchenne muscular dystrophy, hemophilia, fragile X syndrome, and cystic fibrosis. Furthermore, gene sequencing also allows the identification of a person's carrier status with reference to a particular disease and genetic counseling. Another significant application of gene sequencing involves DNA fingerprinting, which involves identifying and establishing an association between biological evidence and suspect during a criminal investigation.

Moreover, gene sequences of individuals can also be stored in a database for any requirement in the future. The US military has kept a database of the genome sequence of all the military personnel so that they can be identified if found missing during an action. Moreover, another application of gene sequencing involves paternity testing to identify the biological parent of a child.

## 5.2   Genetic Engineering

DNA isolation followed by sequencing is widely used to modify plants genetically. This method is widely used in agricultural industries to incorporate DNA containing traits of interest into the plant's genome to give it the desired phenotype. Direct gene manipulation is another way to incorporate changes in the DNA for achieving the required phenotype. A fine example of this is the genetic manipulation of beet crops to gain resistance against herbicide Roundup. Genetic engineering is also used in animals for gene editing and gene cloning.

DNA isolation and sequencing are also used for various recombinant DNA techniques. Two fragments of DNA are joined to prepare new genetic combinations that are then inserted into the host. Another application involves the preparation of recombinant DNA vaccines for inducing immunity against the pathogen (Nascimento & Leite, 2012).

Moreover, genome sequencing has opened new avenues for research in pharmacogenomics to understand an individual's response toward a particular drug. DNA sequencing is also utilized in various epidemiological studies (Di Pietro et al., 2011).

# References

Beutler, E., Gelbart, T., & Kuhl, W. (1990). Interference of heparin with the polymerase chain reaction. *BioTechniques, 9*(2).

Bradley, J., Johnson, D., & Rubenstein, D. (2001). Lecture notes on molecular medicine.

Brawerman, G., Mendecki, J., & Lee, S. Y. (1972). Isolation of mammalian messenger ribonucleic acid. *Biochemistry, 11*(4), 637–641.

Carpi, F. M., Di Pietro, F., Vincenzetti, S., Mignini, F., & Napolioni, V. (2011). Human DNA extraction methods: Patents and applications. *Recent Patents on DNA & Gene Sequences (Discontinued), 5*(1), 1–7.

Nascimento, I. P., & Leite, L. C. C. (2012). Recombinant vaccines and the development of new vaccine strategies. *Brazilian Journal of Medical and Biological Research, 45*, 1102–1111.

# RNA Isolation by the Guanidinium-Acid-Phenol Method

**Abstract** Isolation of RNA by the guanidinium-acid-phenol method is the most commonly used and preferred method in molecular biology. This procedure is also known as the Trizol method of RNA isolation. During this method, the sample is treated with guanidinium isothiocyanate and acidic phenol, and chloroform. Here guanidinium isothiocyanate acts as a powerful protein denaturant and deactivates the RNases, whereas the acidic phenol-chloroform helps in the partitioning of the RNA into the aqueous phase. Finally, the RNA is precipitated out from the aqueous solution using isopropanol and dried well. This method is better as compared to other RNA isolation protocols as the experimenter gets a higher yield of RNA. But this method is slightly time-consuming and uses many harmful chemicals compared to the column-based methods of RNA isolation.

**Keywords** RNA · Phenol · Chloroform · Trizol · RNase · Protein

## 1 Introduction

Guanidinium thiocyanate-phenol-chloroform extraction method is often referred to as a conventional method of RNA isolation. Ulrich et al. 1977 first mentioned guanidinium isothiocyanate in RNA extraction (Cseke et al., 2004). As their method was strenuous, it was soon replaced by a single step and more convenient method called guanidinium thiocyanate-phenol-chloroform extraction (Chomczynski & Sacchi, 2006; Sambrook et al., 1989). Nowadays, these methods are commercialized and are available as kits. Guanidinium thiocyanate and phenol mixture is popularly known as Trizol reagent and is commonly used for RNA extraction. Chloroform is added to the extraction mixture, which then produces a biphasic emulsion. Most of the manufacturers of RNA isolation kits recommend that the samples stored in −80 °C or in liquid nitrogen can also be utilized for extraction if the fresh sample is not available (Sirakov, 2016). Although phenol is toxic and corrosive and denatures protein rapidly, it does not inhibit RNA activity completely (Sambrook et al., 1989).

A. Gautam, *DNA and RNA Isolation Techniques for Non-Experts*, Techniques in Life Science and Biomedicine for the Non-Expert,
https://doi.org/10.1007/978-3-030-94230-4_4

However, the problem was overcome by using phenol-chloroform-isoamyl alcohol (25:24:1), where a biphasic emulsion is formed containing an upper aqueous phase and a lower organic phase. The total RNA, along with some impurities, is contained in the upper aqueous phase. DNA, proteins, lipids, carbohydrates, and cellular debris remain in the emulsion's organic phase (Sambrook et al., 1989). The aqueous phase is collected, and total RNA can be recovered by precipitation with isopropanol.

The yield of total RNA extracted generally ranges from 5 to 10 µg/$10^6$ cells or 4–7 µg/mg tissue. However, it may vary depending on the cell/tissue source. The integrity of extracted RNA is generally verified by agarose gel electrophoresis. This is visualized as discrete bands for rRNA, each for 23S and 16S for prokaryotes, and 18S and 28S for eukaryotes, indicating the other RNA components' intactness. The purity of extracted RNA is calculated in terms of the ratio of $A_{260}/A_{280}$, which generally ranges from 1.8 to 2.0 for good extraction. In common practice, the concentration of RNA is measured at 260 nm by UV spectrophotometry. At this absorption wavelength, 1 absorption unit corresponds to 40 mg/mL of RNA.

# 2   Basic Principle

The guanidinium-acid-phenol method of RNA extraction is mainly based on the phase separation principle of a sample mixture containing a water-saturated phenol component and an aqueous component by centrifugation. Guanidinium thiocyanate being a chaotropic agent, when added to the organic phase of the mixture, leads to the denaturation of proteins, mainly those bound to the nucleic acids and RNAse (Brooks, 1998). After separation, the RNA and some DNA contamination come into an aqueous phase while proteins and other biomolecules along with the cell debris remain in the lower organic phase. RNA is obtained as the precipitate after the addition of isopropanol to the aqueous phase. Then, isopropanol is discarded, washed with ethanol, and air-dried the pellet and dissolved entirely in TE buffer-/RNAse-free deionized water. Other denaturing agents like sarcosine and β-mercaptoethanol aid denaturation (Sambrook et al., 1989; Tan & Yiap, 2009).

# 3   Protocol

## 3.1   Reagents Required

- **Guanidine thiocyanate**
- **0.75 M sodium citrate** (pH 7)
- **10% (w/v) N-lauroyl sarcosinate**
- **Distilled water**

- **14.3 M stock β-mercaptoethanol** (2-ME)
- **Sodium acetate** (anhydrous)
- **Glacial acetic acid**
- **Phenol crystals**
- **49:1 (v/v) chloroform-isoamyl alcohol**
- **Isopropanol**
- **Ethanol**
- **DEPC** (diethylpyrocarbonate)

## 3.2   Working Solutions

| Solutions | Preparation | Storage |
|---|---|---|
| Denaturing solution (4 M guanidine thiocyanate, 25 mM sodium citrate, 0.5% sarkosyl, and 0.1 M 2-ME) | Dissolve 250 g guanidine thiocyanate in 293 mL distilled water. Add 17.6 mL of 0.75 M sodium citrate (pH 7) and 26.4 mL of 10% (w/v) N-lauroyl sarcosinate to it. Mix the solution at ~60 °C. Just before use, add 0.36 mL of 14.3 M stock β-ME per 50 mL of this solution | RT |
| 2 M sodium acetate (pH 4) | Add 16.42 g sodium acetate (anhydrous) to 40 mL distilled water and 35 mL glacial acetic acid. Adjust pH 4 with glacial acetic acid and make up the volume 100 mL with distilled water | RT |
| Water-saturated phenol | Dissolve 100 g phenol crystals in water at 65 °C under the fume hood. Aspirate the upper water phase and collect the lower phase | At 4 °C |
| DEPC-treated water | Add 1 mL of DEPC to 1 L of distilled water in an autoclavable bottle. Keep at room temperature with occasional shaking. Autoclave after 2 h and cool down at room temperature | RT |
| 49:1 (v/v) chloroform-isoamyl alcohol (100 mL) | Mix 98 mL chloroform and 2 mL isoamyl alcohol (3-methyl-1-butanol) in a glass beaker | RT |
| 75% ethanol (100 mL) | Mix 75 mL of absolute alcohol and 25 mL of DEPC-treated water and keep in an airtight bottle | RT |

## 3.3   Procedure

The protocol has been adapted from books on molecular cloning (Buckingham & Flaws, 2007) and short protocol in molecular biology (Ausubel et al., 1992).

1. The critical step in RNA isolation protocol is cell lysis; incomplete lysis would reduce yield and lower purity of isolated RNA. There are different ways to help the process of cell lysis. In the case of tissue cultures, detergents are used. Cell cultures or bacterial cultures can be washed in PBS (phosphate-buffered saline) or physiological saline. In mucous samples like nasal discharges, sputum, and

**Fig. 1** (a) Motorized homogenizer (with attached pestle), and (b) glass-Teflon homogenizer (with pestle) for preparation of the sample homogenate

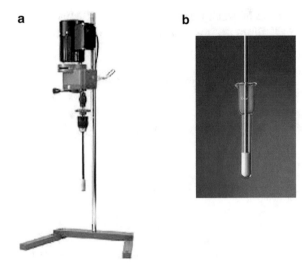

intestinal loops, it is good first to decrease the viscosity of the material (using a mucolytic agent). Gram-positive bacteria are treated with lysozyme; yeasts with zymolyase and paraffin-embedded tissues are treated with xylene to remove the paraffin (Sirakov, 2016).

2. Dissect out the tissue of interest and immediately place it in a clean Petri dish kept on ice.

3. Prepare 10% homogenate of tissue in denaturing solution using a glass-Teflon homogenizer (Fig. 1).

4. Take a 5 mL polypropylene tube, transfer the homogenate to the tube, slowly add the 100 µL sodium acetate (2 M, pH 4), and mix it properly by inverting the tube up and down a few times. Add 1 mL of water-saturated phenol to this mixture. Mix properly, followed by the addition of 200 µL of chloroform-isoamyl alcohol (49:1).

5. Mix it properly by vortexing vigorously for about 10 s and incubate the suspension for 15 min on ice.

6. Centrifuge the tube at 10,000×g for 20 min at 4 °C. Collect the upper aqueous phase and transfer it to the clean tube (Fig. 2). *The upper aqueous phase contains the RNA, whereas the DNA and proteins are in the interphase and lower organic phase. The volume of the aqueous phase is ~1 mL, equal to the initial volume of denaturing solution.*

7. To the aqueous phase collected, add an equal volume of 100% isopropanol for precipitation of RNA and incubate the content at −20 °C for about 1–2 h.

8. After that, centrifuge the content at 10,000×g for 30 min at 4 °C and remove the supernatant.

9. To the pellet obtained, add 300 µL denaturing solution to dissolve the pellet, and transfer to a fresh 1.5 mL microcentrifuge tube.

**Fig. 2**  Different layers in a microfuge tube obtained during RNA extraction by the GITC method

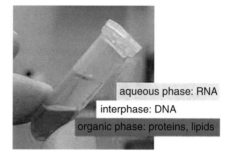

**Fig. 3**  1% agarose gel showing different types of ribosomal RNA in control and experimental samples

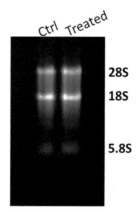

10. Again, add 300 μL (equal volume) of isopropanol, incubate for 1 h at −20 °C and centrifuge at 10,000×g for 10 min at 4 °C and remove the supernatant.
11. Wash the pellet in 75% ethanol for 10–15 min at RT and air-dry the RNA pellet.
12. Add 20–100 μL DEPC treated water or TE buffer and dissolve the RNA pellet properly.
13. Store the dissolved RNA at −80 °C till further use.
14. Quantitate RNA and check its integrity on the agarose gel (Fig. 3).

# 4   Precautions

- Phenol is highly flammable and corrosive. So, use gloves and keep proper safety measures for eye protection while working with phenol.
- Use of water-saturated phenol is recommended rather than buffered phenol.
- Use talc-free gloves and nuclease-free plastic wares.
- Do not overdry the RNA pellet as the hardened pellet will not dissolve properly.

## 5    Applications

- High-quality RNA is an essential requirement to get good and reliable results in several biochemical and molecular techniques like qRT-PCR, cDNA library construction, northern blotting, transcriptome studies, array analysis, etc. (Buckingham & Flaws, 2007; Sirakov, 2016).
- Detection of even any trace of nucleic acids is done in different areas such as diagnostics of various infectious diseases and genetic disorders, marker-assisted selection or assessment of variation in biodiversity, etc. (Doyle, 1996; Sirakov, 2016).
- Several methods have been developed for nucleic acid detection based on RNA's hybridization, such as in situ hybridization, molecular beacon, and polymerase chain reactions (PCR, reverse transcription PCR, real-time PCR) (Buckingham & Flaws, 2007).

## References

Ausubel, F. M., Brent, R., Kingston, R. E., et al. (1992). *Short protocols in molecular biology* (2nd ed.). Green publishing associates and John Wiley and sons.

Brooks, G. (1998). *Biotechnology in healthcare: An introduction to biopharmaceuticals.* Pharmaceutical Press.

Buckingham, L., & Flaws, M. L. (2007). *Molecular diagnostics: Fundamentals, methods, & clinical applications.* F.A. Davis.

Chomczynski, P., & Sacchi, N. (2006). The single-step method of RNA isolation by acid guanidinium thiocyanate-phenol-chloroform extraction: Twenty-something years on. *Nature Protocols, 1*(2), 581–585.

Cseke, L. J., Kaufman, P. B., Podila, G. K., & Tsai, C. J. (2004). *Handbook of molecular and cellular methods in biology and medicine* (2nd ed.). CRC Press.

Doyle, K. (1996). *The source of discovery: Protocols and applications guide.* PROMEGA.

Sambrook, J., Fritsch, E. F., & dan Maniatis, T. (1989). *Molecular cloning a laboratory manual* (2nd ed.). Cold Spring Harbor Laboratory Press.

Sirakov, I. N. (2016). *Nucleic acid isolation and downstream applications, nucleic acids—From basic aspects to laboratory tools, Marcelo L. Larramendy and Sonia Soloneski.* IntechOpen. https://doi.org/10.5772/61833

Tan, S. C., & Yiap, B. C. (2009). *DNA, RNA, and protein extraction: The past and the present journal of biomedicine and biotechnology.* Hindawi Publishing Corporation. https://doi.org/10.1155/2009/574398

# Spin Column-Based Isolation of Nucleic Acid

**Abstract** Spin column technique is a solid-phase extraction commercial strategy to extract nucleic acid from a wide range of crude biological samples, including tissues, plant extracts, viruses, and bacteria. It is advantageous over other extraction techniques such as less time consumption, economical, energy-efficient, automated operation, prevention from cross-contamination, and high yield. It is based on the principle of binding nucleic acid to immobilized solid-phased spin columns of different materials under specific circumstances. There are five commercial types of spin column used for nucleic acid extraction, including silica membrane, anion exchange, filter paper, glass fiber, and polyethylene fibers. It is a five-stage process consisting of cell lysis, purification, washing, dry spin, and elution using appropriate buffers. This technique possesses applications in molecular studies, diagnosis, forensic science, vaccine development, and pharmaceuticals. The techniques differ for DNA and RNA extraction in maintaining the pH of elution buffer (basic for DNA), which is the most crucial stage of separation processing. However, the automated operation of spin columns is required to connect with the Internet of things to develop a next-generation advanced intelligent separation system.

**Keywords** Spin column · Elution · Nucleic acid · Cell lysis · Anion exchange · Silica membranes · Solid-phase extraction

## 1 Introduction

The potential of spin columns for nucleic acid extraction and purification was recognized in the 1980s (Marko, 1982). It is a solid-phase extraction technique in which the target molecules (nucleic acids) bind to immobilized solid-phased silica resin under specific conditions. The use of spin columns for nucleic acid isolation possesses numerous advantages over other techniques, including high yield, cost-effectiveness, automated processing, less time consumption, environment friendly, adaptability, and significant scalability (Tolosa, 2007). Furthermore, this technique

© The Author(s), under exclusive license to Springer Nature Switzerland AG 2022　　47
A. Gautam, *DNA and RNA Isolation Techniques for Non-Experts*, Techniques in
Life Science and Biomedicine for the Non-Expert,
https://doi.org/10.1007/978-3-030-94230-4_5

also addresses the challenge of cross-contamination by excluding hazardous chemicals used in conventional nucleic acid extraction methods. The fascinating feature of this technique is that spin column separation is a simple, reliable, and quick technique to simultaneously extract RNA, DNA, and proteins from a single biological sample, saving time and money and allowing proficient use of small biological samples (Tan, 2009). The apparatus required is exceedingly cost-effective both in terms of manufacturing and being installed in laboratory settings. This extraction technique can be used to extract nucleic acids from a wide range of crude biological samples, including blood, tissues, plant parts and extracts, viruses, and bacteria. The extracted nucleic acid can be used for diversified downstream applications in gene expression analysis, biomedical research, and molecular diagnostics (Dash, 2020).


## 2   Basic Principle

Spin columns are the apparatus that allows the commercial extraction and purification of nucleic acids in molecular biology laboratories. Different filters with distinctive bore sizes, such as 2/3/4/6/8 layers, are used in spin columns to specifically target nucleic acid or its part (such as plasmid RNA, viral DNA/RNA, genomic DNA, DNA fragments) (Dash, 2020). There are majorly five different types of spin columns that are widely used to extract nucleic acids, which have been mentioned below:

**Silica membrane-based spin column**: The use of silica membrane as gel material in the spin column is the most popular and widely used technique (Husakova, 2020). DNA or RNA binds to silica by adsorption. The use of chaotropic agents (like guanidinium HCl) denatures and reduces the hydrophobicity of biomolecules. In the presence of a high salt concentration, chaotrope allows the formation of a positive salt bridge between negatively charged silica and negatively charged nucleic acid molecules. In the silica spin column, silica is bound to the solid support, which addresses the challenge of glass-bead contamination of extracted nucleic acids and shearing of DNA fragments during extraction, which lies or exceeds the range of 3–10 kb.

**Anion exchange-based spin column**: Most transfection-grade plasmid DNA extraction and purification can be done using this technique (Chang, 2008). During anion exchange-based spin column isolation of DNA/RNA, a positively charged resin (like diethylaminoethyl cellulose or DEAE cellulose) is used in the column to attract and bind the negatively charged phosphate groups of the nucleic acid molecules. These days membrane-based spin architecture eliminates column packing resulting in quicker extraction. The centrifuge format enables the handling of many samples simultaneously.

**Filter paper-based spin column**: In recent years, cellulose fiber-based filter paper has emerged as a substitute for silica material in spin columns (Shi, 2018). Though it can extract long and linear ds DNA from most crude samples, filter paper shows weak binding affinity to the plasmid DNA. The basic principle of DNA/RNA isolation is similar to that of silica membrane-based isolation in the presence of chaotropic agents. Here, the secondary fibril-associated cellulose of filter paper binds to the nucleic acids in chaotropic conditions. It is a cost-effective, readily available, and less time-consuming technique for extracting and purifying nucleic acids.

**Glass fiber-based spin column**: Glass fiber spin columns can be used for the purification of both genomic RNA and plasmid DNA (Ali, 2017). During the extraction/purification process, it holds nucleic acids with negligible loss and eradicates slats and small molecule contamination. It results in a high yield of nucleic acids with significant purity. The silica or derivatized silica of the glass fiber binds to the nucleic acid molecules in the presence of chaotropic salts, similar to the mechanism mentioned above.

**Polyethylene (PE) fiber-based spin column**: PE fibers are significantly chemically inherent, thermally stable, and hydrophobic (Dash, 2020). Thus, they can withstand a wide range of buffer solutions and samples. Polyethylene fibers bind to DNA covalently, which can be eluted later on using suitable elution buffers. PE fiber spin columns are used for coarse filtration and fine particle removal. This method is widely used as a pre-filtration step before the extraction/purification process.

# 3   Protocol

Since the suppliers of columns customize the protocols and reagents, a general idea of the whole procedure has been mentioned below.

## 3.1   Reagents Required

- Sodium dodecyl sulfate (SDS)
- Sodium chloride (NaCl)
- Spin columns
- Ethanol (absolute alcohol)
- Deionized water
- Tris-base (pH 10–11)
- Acidic water (pH 4–5)
- Chaotropic salts (urea, guanidine thiocyanate, guanidine HCL, lithium perchlorate)
- Enzyme: Proteinase K or lysozyme

## 3.2   Working Solution

| Solution | Preparation |
| --- | --- |
| Lytic buffer | Chaotropic salt + detergent + enzyme |
| Purifying buffer | 1.25 M NaCl + ethanol + deionized water |
| Washing buffer | Low concentration chaotropic salt  thanol |
| Elution buffer | 10 mM Tris base (pH 8–9) for DNA and water (pH 4–5) for RNA |

## 3.3   Procedure

The process of spin column-based extraction can be broadly divided into five stages (Fig. 1).

**Cell lysis**: In the first stage, the samples are treated with lytic buffers to break the cell membrane and remove nucleic acid. These lytic buffers consist of a high concentration of chaotropic salts, a detergent, and an enzyme. Various chaotropic salts such as urea, guanidine thiocyanate, guanidine HCl, and lithium perchlorate are used in the first stage. It plays two significant roles in nucleic acid extraction: destabilization of proteins (by destabilizing hydrogen, van der Waals, or hydrophobic interactions) and disruption of the association of nucleic acid with water (to transfer to silica). A detergent (SDS) is also used in the lytic buffer to assist cell lysis and protein solubilization through its micellar action. Enzymes such as Proteinase K and lysozyme are efficiently used during the first stage, depending upon the sample type. Proteinase K digests proteins away from nucleic acid preparations in denaturing lytic buffer. However, lysozyme is used before denaturing lytic buffer, as it becomes inactive in the lytic buffer. Hence, the addition of lytic buffer releases the nucleic acid by breaking the cell membrane and assists in the transfer of nucleic acid to silica/another column.

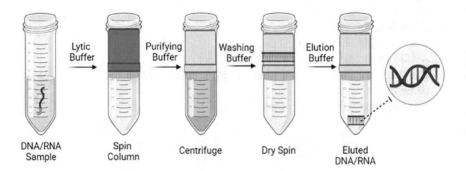

**Fig. 1** Spin column-based nucleic acid extraction process

**Purification**: In the second stage, the optimal amount of ethanol or isopropanol is added to further accelerate the binding of nucleic acids to the column. The amount of ethanol used is of prime importance as a high amount of ethanol can cause degradation, and its lower amount impedes the washing of the salt from the membrane. Additionally, sometimes sodium chloride is used as a precipitant to evade contamination due to detergent in the first stage.

**Washing**: In the third stage, the sample is first centrifuged. It results in the binding of desired nucleic acids to the column's silica membrane, and the impurities, including polysaccharides and proteins, remain in flow-through. However, some proteins and salt residues also stick to the silica membrane. Such impurities are removed through washing, which is a two-step washing procedure. In the first step, washing is done by using a low concentration of chaotropic salts to remove remnant proteins and pigments. This step is followed by two-time ethanol washing to remove remnant chaotropic salts.

**Dry spin**: In the fourth stage, residual ethanol is washed from a column containing nucleic acids. This is essential to get a clean eluent, as the residual ethanol prevents the hydration of nucleic acid during elution. Thus, a dry spin run of a column containing nucleic acid eliminates residual ethanol for better elution.

**Elution**: In the final stage, the spin column is added with elution buffer and centrifuged, which removes the nucleic acid from the column. Further, the nucleic acid is collected from the base of the column. However, there is a slight difference in the elution of DNA and RNA.

A slightly basic pH elution buffer is used for DNA extraction as DNA is stable in a basic pH medium and dissolves in it faster than water. Thus, a basic buffer comprised of 10 mM Tris at pH 8–9 is generally used. This is also done in the case of DNA pellets. In a typical process, a column containing DNA is first placed in an elution buffer for a few minutes and then centrifuged. This is done because high molecular weight DNAs require more time to rehydrate than DNA with less molecular weight.

However, for RNA extraction, water is used as eluent as water pH is 4–5. This is because nucleic acid is more stable and dissolves easily in slightly acidic pH.

# 4   Precautions

The following precautionary measures must be taken to get a high yield of DNA/RNA:

- Use high-quality ethanol (100%, 200 proof) throughout the process, since the poor-quality ethanol predominantly affects the DNA/RNA yield due to water and other contaminants.
- Properly wash the samples removing all the remnants, including salts, proteins, and pigments.

- Use an optimized concentration of ethanol, as a higher/lower concentration of ethanol during extraction may cause a low or contaminated yield of DNA/RNA.
- A dry run of the sample in stage four is mandatory as it removes residual ethanol in silica membranes, which prevents rehydration of DNA/RNA. It is generally missing in most protocols, which is a possible reason for low and contaminated yield.
- Maintain the pH of elution buffer according to the desired nucleic acid, which is to be extracted.

## 5 Applications

Molecular studies: Extracted DNA/RNA is used for various genomic and hereditary studies, including cloning, replication, and sequencing (Lee, 2020) (Dash, 2020).

Diagnostics: Various medical conditions such as hemophilia A, sickle-cell anemia, Down's syndrome, cystic fibrosis, and fragile X syndrome can be diagnosed early through the help of extracted DNA/RNA. It also possesses importance in virology due to its use in the detection of diseases like COVID-19. It not only helps diagnose diseases but also helps in speculating whether a patient is a potential carrier of it or not (Ali, 2017) (Smyrlaki, 2020).

Pharmaceuticals and vaccines: The extraction of nucleic acids is the initial step in researching and developing various pharmaceuticals and vaccines. Recombinant genetic engineering has helped to design pharmaceuticals based on insulin and human growth hormone. It is of prime importance in vaccine development as nucleic acid-based vaccines are considered potentially safer than conventional vaccines, especially in the case of COVID-19, malaria, and hepatitis (Wommer, 2021) (Ali, 2017) (Tan, 2009).

Forensic science: Nucleic acid extraction is a key technique in forensic sciences for different purposes, including fingerprinting, identifying the criminals by matching the DNA of the accused obtained from the crime site, and identifying parenting (Tolosa, 2007) (Dierig, 2020).

## References

Ali, N. R. (2017). Current nucleic acid extraction methods and their implications to point-of-care diagnostics. *BioMed Research International, 9306564.*

Chang, C. N. (2008). Preparation of inorganic-organic anion-exchange membranes and their application in plasmid DNA and RNA separation. *Journal of Membrane Science, 311*(1–2), 336–348.

Dash, H. S. (2020). Isolation of DNA by using column-based extraction system. In *Principles and practices of DNA analysis: A laboratory manual for forensic DNA typing.* Springer Protocols Handbooks.

Dierig, L. S. (2020). Looking for the pinpoint: Optimizing identification, recovery and DNA extraction of micro traces in forensic casework. *Forensic Science International: Genetics, 44*, 102191.

Husakova, M. K. (2020). Efficiency of DNA isolation methods based on silica columns and magnetic separation tested for the detection of Mycobacterium avium Subsp. *Paratuberculosis in Milk and Faeces. Materials, 13*(22), 5112.

Lee, K. T. (2020). Parallel DNA extraction from whole blood for rapid sample generation in genetic epidemiological studies. *Frontiers in Genetics, 11*, 374.

Marko, M. C. (1982). A procedure for the large-scale isolation of highly purified plasmid DNA using alkaline extraction and binding to glass powder. *Analytical Biochemistry, 121*(2), 382–387.

Shi, R. L. (2018). Filter paper-based spin column method for cost-efficient DNA or RNA purification. *PLoS One, 13*(12), e0203011.

Smyrlaki, I. E. (2020). Massive and rapid COVID-19 testing is feasible by extraction-free SARS-CoV-2 RT-PCR. *Nature Communications, 11*, 4812.

Tan, S. Y. (2009). DNA, RNA, and protein extraction: The past and the present. *BioMed Research International, 2009*, 574398.

Tolosa, J. S. (2007). Column-based method to simultaneously extract DNA, RNA, and proteins from the same sample. *BioTechniques, 43*(6), 799–804.

Wommer, L. M. (2021). Development of a 3D-printed single-use separation chamber for use in mRNA-based vaccine production with magnetic microparticles. *Engineering in Life Sciences*, 1–16.

# Isolation of Plasmid DNA by Alkaline Lysis

**Abstract** Apart from the chromosomal DNA, many microorganisms contain plasmid DNA too. This plasmid is a small circular DNA that can replicate independently of the chromosomal DNA. They provide antibiotic resistance and hence better adaptation qualities to these microorganisms to survive in different environmental conditions. Molecular biologists harness these properties of plasmid DNA and use them for several applications such as sequencing, screening clones, transfection, PCR, cloning, and restriction digestion. Therefore, the isolation of plasmids from cells is required during such applications. Isolation of plasmid DNA by alkaline lysis method involves using suitable biochemicals at standardized pH and concentration to selectively precipitate the circular DNA out of the noncircular chromosomal DNA. After digesting it with a suitable restriction enzyme, the extracted plasmid DNA can be qualitatively checked by agarose gel electrophoresis.

**Keywords** Plasmid · Circular DNA · Extrachromosomal DNA · Alkaline lysis · Recombinant DNA

## 1 Introduction

Bacterial plasmid DNAs are extensively used as cloning vectors in research involving recombinant DNA techniques. Plasmids are self-replicating, circular, double-stranded extrachromosomal DNA molecules (Fig. 1). Once a new plasmid is constructed, it may be isolated and characterized for its size and restriction enzyme pattern by gel electrophoresis. Plasmid DNA may be isolated from small-scale bacterial cultures by treatment with alkali and SDS. The resulting DNA preparation may be screened by electrophoresis or restriction endonuclease digestion. With further purification, the preparations may be used as templates in DNA sequencing reactions. Birnboim and Doly first described the alkaline lysis method of plasmid DNA isolation in 1979 (Birnboim & Doly, 1979; Birnboim, 1983). Since then, it is the most preferred method of plasmid DNA isolation with a few modifications.

© The Author(s), under exclusive license to Springer Nature Switzerland AG 2022
A. Gautam, *DNA and RNA Isolation Techniques for Non-Experts*, Techniques in
Life Science and Biomedicine for the Non-Expert,
https://doi.org/10.1007/978-3-030-94230-4_6

**Fig. 1** Plasmid (pUC19)
showing the origin of
replication, drug-resistant
and multiple cloning sites
(MCS)

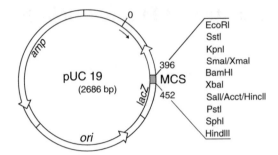

Specific protocols for alkaline lysis differ from laboratory to laboratory; however, they are all based on the same principle. The first stage is to grow the selected bacterial colonies in a small volume of lysogeny broth (LB) media containing the selection antibiotic and then process them for alkaline lysis, more popularly known as the miniprep method of plasmid isolation.

Using this procedure, 2–5 µg of DNA can be obtained from a 1.5 mL culture of *E. coli* containing a pBR322-derived plasmid, and three- to fivefold higher yields can be expected from pUC-derived plasmids (Ausubel et al., 1992). The success of using the alkaline lysis method mainly depends on the strain of *E. coli* used. Strains having high endonuclease activity, such as HB101 or the JM100 series, yield DNA that often requires further purification with a phenol-chloroform extraction or through additional methods of precipitation (Ausubel et al., 1992; Sambrook et al., 1989). However, the alkaline lysis procedure seems to be the most used plasmid purification protocol regardless of the strain. It is also better suited for isolation of high molecular weight (>10 kb) or low copy number plasmids than the boiling lysis method. Plasmid DNA isolated by alkaline lysis is suitable for most analyses and cloning procedures without further purification. The isolated covalently closed circular (CCC) plasmid DNA can be resolved by agarose gel electrophoresis either intact or after digesting it with a suitable restriction enzyme and may be examined further.

## 2   Basic Principle

Earlier works have shown that denaturation of linear DNA but not CCC-DNA occurs at a narrow range of pH (about 12.0–12.5). This property is exploited for purifying CCC-DNA (Birnboim, 1983; Birnboim & Doly, 1979). Isolation of plasmid from the chromosomal DNA of bacteria is based on their differential denaturation property during the alkaline lysis. During this treatment, chromosomal as well as plasmid DNA is denatured. By carefully selecting the ratio of the cell suspension to NaOH solution, an alkaline pH value is obtained; further pH control is obtained by including glucose as a pH buffer. This buffer is often a basic pH Tris buffer, which helps denature DNA and EDTA (ethylenediaminetetraacetic acid) that binds divalent cations destabilizing the membrane and inhibiting DNases (enzymes that degrade DNA). The SDS detergent dissolves the cell membrane's phospholipids and proteins, resulting in cell lysis and the release of the cellular components. The high concentration of sodium hydroxide denatures the genomic and plasmid DNA, as well as cellular proteins. The cellular DNA becomes linear, and the DNA strands are separated, whereas the plasmid DNA is circular and remains topologically constrained (the two strands, although denatured, remain together). Subsequent neutralization with potassium acetate allows only the covalently closed plasmid DNA to reanneal and to stay solubilized. At the same time, the chromosomal DNA, still in a very high molecular weight, precipitates in a complex formed with potassium and SDS, which is removed by centrifugation (Casali & Preston, 2003). Simultaneously, the high concentration of potassium acetate causes precipitation of protein-SDS complexes and high molecular weight RNA (Birnboim & Doly, 1979). By doing this, the three major contaminating molecules are coprecipitated and can be removed by single centrifugation. Plasmid DNA is recovered from the supernatant by ethanol precipitation.

# 3   Protocol

The protocol has been adapted from Sambrook et al. (1989).

## 3.1   Reagents Required

- **1 M glucose (50 mL)**: Mix 9 g glucose in 50 mL distilled water.
- **1 M Tris-Cl (pH 8, 50 mL)**: Mix 6.057 g Tris base in 45 mL distilled water. Maintain pH by HCl and make up volume to 50 mL by adding more water.
- **0.5 M EDTA (pH 8, 100 mL)**: Mix 14.61 g EDTA in 95 mL distilled water. Maintain pH by NaOH and make up volume to 100 mL by adding more water.
- **10 N NaOH (50 mL)**: Add 20 g NaOH pellets in 30 mL distilled water. Make up volume to 50 mL by adding more water.
- **1% (w/v) SDS (50 mL)**: Add 0.5 g SDS in 50 mL distilled water.
- **5 M potassium acetate (100 mL)**: Add 49 g potassium acetate to 100 mL distilled water.
- **Glacial acetic acid**.
- **Distilled (deionized) water**.
- **Ethanol**.
- **Isopropanol**.

## 3.2   Working Solution

| Solution | Preparation | Storage |
|---|---|---|
| Lysis buffer (alkaline lysis solution I) (100 mL) | Mix 5 mL of 1 M glucose, 2.5 mL of 1 M Tris-HCl, and 1 mL of 0.5 M EDTA in 90.5 mL of distilled water | RT |
| Denaturing solution (alkaline lysis solution II) (100 mL) | Mix 0.2 mL of 0.2 NaOH and 1 mL of 1% SDS in 8.8 mL of distilled water | RT |
| Neutralizing solution (alkaline lysis solution III) (100 mL) | Mix 60 mL of 5 M potassium acetate and 11.5 mL of glacial acetic acid in 28.5 mL of distilled water | RT |
| 70% ethanol (100 mL) | Add 30 mL distilled water in 70 mL absolute alcohol (ethanol) | RT |
| Isopropanol | Use as purchased | RT |
| TE buffer (pH 8) | Mix 10 mL 1 M Tris buffer and 1 mL 0.5 M EDTA in 75 mL distilled water. Maintain pH 8 and make up volume to 100 mL by adding more water | RT |

## 3.3   Procedure

1. Pour overnight grown culture of bacteria into a 1.5 mL microcentrifuge tube.
2. Centrifuge the culture suspension at ~12,000×$g$ for 1 min at 4 °C.
3. Discard the whole supernatant carefully without disturbing the bacterial pellet, leaving it as dry as possible.
4. Resuspend the pellet in 100 μL of ice-cold alkaline lysis solution I and vortex it properly.
5. Add 200 μL of freshly prepared lysis solution II to this suspension, and mix it properly by inverting the tube several times.
6. Then, add 150 μL of alkaline lysis solution III to the mixture, and mix the whole suspension thoroughly by inverting the tube up and down several times and putting it on the ice for about 5 min.
7. Centrifuge the mixture at 14,000×$g$ for 5 min at 4 °C, and transfer the supernatant to a fresh tube.
8. To the obtained supernatant, add absolute ethanol twice the volume of supernatant, and mix it well and allow it to stand for at least 5 min at RT.
9. Then, centrifuge it for 3 min at RT, discard the supernatant, and wash the pellet with 1 mL of 70% ethanol followed by air-drying the pellet.
10. Resuspend the air-dried pellet in 30–50 μL of TE buffer and 1 μL of 10 mg/mL RNase solution.
11. Dissolve it by gentle vortexing for a few seconds.
12. This solution contains isolated plasmid DNA. Store it at −20 °C till further use.
13. Analysis of isolated DNA:

**Fig. 2** Resolved plasmid
DNA on agarose gel,
showing the relative
position of different types
of plasmid bands after
electrophoresis

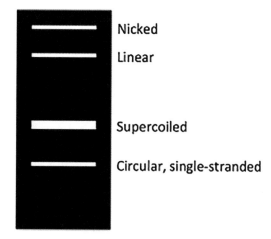

**Fig. 3** Spectrophotometer

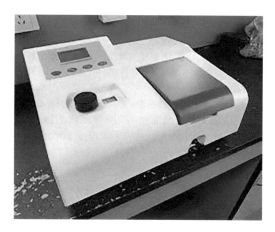

### 3.3.1  Agarose Gel Electrophoresis

The isolated plasmid DNA, preferably uncut DNA, is analyzed by running it on 1%
agarose gel, which is likely to produce three bands. Also, this is the way to verify
the integrity of isolated plasmid DNA and check whether there is any contamination
of chromosomal DNA (Fig. 2). In addition, plasmid DNA can be analyzed by cleavage with restriction enzymes.

### 3.3.2  Determination of Plasmid Concentration by Spectrophotometry

The concentration of plasmid DNA isolated from the given bacterial culture can be
quantified using a spectrophotometer (Fig. 3). Any chromosomal DNA and/or RNA
contamination also contributes to the concentration readings, thus giving a possible

false measurement. Therefore, the optical density (O.D.), i.e., absorbance, is taken at 260 nm and 280 nm to overcome this issue. The ratio A260/A280 tells about the purity of the preparation. A value of $1.8 \pm 0.05$ for the A260/A280 ratio is an accepted purity level.

The concentration of DNA can be calculated using the formula given below:

DNA ($\mu$g/mL) = A260 $\times$ Dilution factor $\times$ 50

(Here, 50 is the extinction coefficient of DNA.)


# 4   Precautions

- The bacterial pellet should be completely free from the growth medium. Otherwise, any presence of the growth medium, possibly due to insufficient washing of the DNA, may cause interference in the cleavage of plasmid DNA by restriction enzymes.
- The final pellet of DNA should not be overdried; otherwise, it becomes challenging to dissolve.
- In general, it is not mandatory to add lysozyme for isolation of the plasmid DNA. However, when lysozyme is added to the GTE solution, it gives enhanced DNA yield.
- Nucleic acid preparation should not be diluted beyond a critical limit to produce reliable DNA concentration measurement. Commonly, the samples are diluted to a limit where it gives a minimum absorbance of 0.1.


# 5   Applications

**Plasmids in gene therapy**: Owing to their ease of manipulation and replication in the bacterial cell, plasmids are used to insert therapeutic genes into the human body to fight against diseases (Fig. 3). As compared to viral vectors, plasmids are with minimum or no harm. Plasmids are commonly adopted in the life sciences as vectors for introducing foreign DNA into another cell.

**Plasmids in recombinant DNA technology and genetic engineering**: Plasmids are more commonly used in genetic engineering, for example, to amplify a given gene or set of genes or to insert a gene or drug of interest using plasmid DNA. Artificial and cost-effective bulk production of antibiotics is achieved through the use of plasmids in industries. Also, the plasmid DNA, which is synthesized artificially, is being utilized in several important fields such as the generation of genetically modified plant species, therapeutic drugs and many proteins, etc.

# References

Ausubel, F. M., Brent, R., Kingston, R. E., et al. (1992). *Short protocols in molecular biology* (2nd ed.). Green publishing associates and John Wiley and sons.

Birnboim, H. C. (1983). A rapid alkaline extraction method for the isolation of plasmid DNA. *Methods in Enzymology, 100*, 243–255.

Birnboim, H. C., & Doly, J. (1979). A rapid alkaline extraction procedure for screening recombinant plasmid DNA. *Nucleic Acids Research, 7*(6), 1513–1523.

Casali, N., & Preston, A. (2003). *E. coli plasmid vectors : Methods and applications methods in molecular biology; v.* Humana Press Inc.

Sambrook, J., Fritsch, E. F., & dan Maniatis, T. (1989). *Molecular cloning a laboratory manual* (2nd ed.). Cold Spring Harbor Laboratory Press.

# DNA Isolation by Hydrophilic Ionic Liquid Treatment

**Abstract** For quick and quantitative DNA extraction, hydrophilic ionic liquids (ILs) are highly beneficial. This DNA extraction procedure is affordable and takes only a few minutes to complete. Ionic liquids can dissolve biomass and release DNA molecules from tissues and cells. DNA released in the supernatant may be used as such for downstream applications. The extraction procedure includes mixing ionic liquids with starting sample at a high temperature (65–95 °C), depending upon the sample source. If the plant or animal tissues are being used, brief centrifugation may be needed. Aqueous supernatant may be directly used for downstream applications. This approach is suited for high sample throughput and allows DNA extraction from bacteria, plant, and animal tissues in resource-constrained environments. Therefore, this method is most suitable for the detection of contaminations in food. The major disadvantage of this approach is that it might result in highly diluted DNA, which requires subsequent concentration by using DNA-binding columns or silica-coated magnetic beads.

**Keywords** Ionic liquids · qPCR · Choline hexanoate · PCR enhancement · Food safety

## 1 Introduction

DNA is one of the most studied biomolecules because of its central role in the function of all living organisms. DNA extraction by hydrophilic ionic liquids (ILs) is a new method, which is very useful for fast and quantitative DNA extraction, followed by molecular biological techniques such as quantitative polymerase chain reaction (qPCR). This technique of DNA extraction is inexpensive, uses no toxic chemicals, and takes only 5 min to complete. This approach is suited for high sample throughput and allows DNA extraction from bacteria, plant, and animal tissues in resource-constrained (Fuchs-Telka et al., 2016; Gonzalez García et al., 2014).

© The Author(s), under exclusive license to Springer Nature Switzerland AG 2022    63
A. Gautam, *DNA and RNA Isolation Techniques for Non-Experts*, Techniques in
Life Science and Biomedicine for the Non-Expert,
https://doi.org/10.1007/978-3-030-94230-4_7

DNA is negatively charged, with phosphate groups, a pentose sugar, and nitrogenous bases forming its double-helical structure. The two strands of DNA connect through a hydrogen bond. The nitrogenous bases exert a hydrophobic effect, while phosphate and sugar groups are hydrophilic, oriented toward the exterior side and interacting with water. Ionic liquids stabilize DNA by forming electrostatic bonds between negatively charged phosphate groups and their cations, although a higher concentration of ionic liquids causes the destabilization of DNA. Besides the electrostatic interaction, groove binding of ionic liquid cation through hydrophobic and polar interaction increases DNA stability (Shukla & Mikkola, 2020). Ionic liquids have been discovered to be superior to water and other volatile organic solvents in terms of stabilizing or disrupting the 3-D structure of DNA or protein. In addition, the charge density of ionic liquids allows them to be hydrophilic or hydrophobic, retaining their properties over a wide temperature range.

## 2   Basic Principle

Ionic liquids have been frequently utilized as solvents because of their advantages over volatile solvents. Their single cation can couple with numerous anions and vice versa, allowing the formation of zwitterions. These salts exist as liquids below 100 °C and are typically made up of a mix of organic cations and organic/inorganic anions (Shamshina et al., 2019). Because they have both directional and nondirectional forces, they can easily interact with a broad spectrum of polar and nonpolar solutes, including biomolecules like proteins and DNA (Fig. 1). New methods employing ionic liquids have been effectively utilized in DNA extraction. Ionic liquids can dissolve biomass and release DNA molecules from tissues and cells. Further, DNA released in the supernatant may be used as such for downstream applications. It can be stored in ionic liquids at room temperature because it also preserves DNA integrity in the presence of nucleases. Therefore, ionic liquids provide a handy and quick method for DNA extraction without using sophisticated instruments and hazardous chemicals.

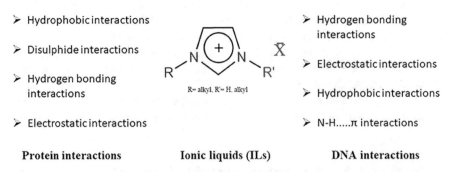

**Fig. 1** Interactions of ILs with protein and DNA. (Adapted from Shukla & Mikkola, 2020)

# 3   Protocol

This protocol is adapted from Martzy et al. (2019), using a hydrophilic ionic liquid choline hexanoate ([Cho]Hex in Tris pH 8 buffer).

## 3.1   Reagents Required

- Choline hydroxide (0.1 M)
- Hexanoic acid (0.1 M)
- Tris pH 8 buffer
- Bacterial cell suspension

## 3.2   Working Solutions

| Solution | Preparation | Storage |
|---|---|---|
| 50% w/w [Cho]Hex in Tris pH 8 buffer | Prepare a fresh solution of choline hydroxide (0.1 M) in a round bottom flask by adding distilled water. Add hexanoic acid (0.1 M) dropwise to neutralize the solution. Stir the reaction mixture at room temperature continuously using a magnetic stirrer for 12 h. Finally, concentrate in a vacuum evaporator. A light yellowish gel will be obtained after spin-dry for 20 h. Mix the ionic liquid and Tris-Cl buffer in a 1:2 ratio | Room temperature (RT) |
| Tris buffer (10 mM pH 8.0– 100 mL | Add 10 mL of stock solution to 90 mL of deionized water | RT |

## 3.3   Procedure

### 3.3.1   Procedure of DNA Extraction

1. Centrifuge the cell homogenate or suspension of log-phase cultures of bacterial cells (OD670 = 0.2) at 10,000×$g$ for 5 minutes.
2. Resuspend the pellets into Tris buffer (10 mM, pH 8).
3. Use 10 μL of respective cell suspensions for the extraction procedure.
4. Mix 10 μL of resuspended with 90 μL IL/buffer system ([Cho]Hex in Tris pH 8 buffer), and incubate at 65 °C for 5 min.

Note: 1. *If starting material is plant tissue, stir 100 mg of tissue in 900 mg of an IL for 1 h at room temperature. Dilute the homogenate with 5 mL of deionized water and incubate at 95 °C for 10 min.*

*2. If starting material is animal tissue, dissolve 100 mg of IL in 900 mL of Tris buffer (10 mM, pH 8.0). Add 200 mg of the tissue sample to this solution, and stir for 15 min at room temperature. Incubate the whole mixture at 95 °C for 10 min.*

5. Cell debris will be settled down, and the upper aqueous layer contains DNA which can be directly used for PCR amplification without further purification (Fig. 2).

*Note: If the starting material is a plant or animal tissue, brief centrifugation for 5 minutes at 10,000×g is recommended. The supernatant is utilized for further amplification.*

# 4   Precautions

- If the targeted DNA is diluted below the detection limit of the corresponding analytical technique, it may result in false-negative results. To circumvent this restriction, DNA-binding columns or silica-coated magnetic beads might be employed to separate nucleic acids from the crude extract. To overcome inhibi-

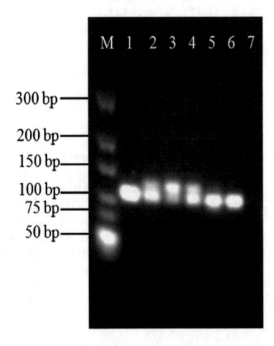

**Fig. 2** DNA isolated from beef extracts with different ILs (*1*) Choline formate, (*2*) choline acetate, (*3*) choline lactate, (*4*) choline butyrate, (*5*) choline hexanoate, (*6*) choline octanoate, was amplified by PCR and separated on agarose gel. (Adapted from Ressmann et al., 2015)

tory effects, dilute the extract 1:20 with 10 mM Tris buffer before subsequent qPCR analysis.
- It is essential to handle organic materials carefully because they are sometimes toxic and harmful to the skin, so always use hand gloves, wear a lab coat, and wear safety glasses while handling the organic products.
- The tubes must be properly balanced during centrifugation.

# 5  Applications

Though the isolated DNA can be used as an input for all molecular studies, two major applications have been listed here.

## 5.1  Food Safety Surveillance

Food composition and authenticity are major concerns for food regulators. Moreover, manufacturing processes may not always comply with high manufacturing standards, and traceability of the ingredients' origin may be missing. Furthermore, detection of genetically modified food products has become increasingly essential in recent years. Laboratories must be able to determine the composition of processed foods and meats to protect customers against mislabeled products.

The initial stage in the screening procedure is to extract the DNA so that the species-specific sequences may be amplified later. DNA extraction from meat and processed food materials must be handy and robust, without any requirement of sophisticated instruments. Therefore, ionic liquids are good for DNA extraction because of their intrinsic features, such as their capacity to dissolve biomass, DNA stability in hydrated ionic liquids, and PCR enhancement.

## 5.2  PCR Enhancement

DNA can be stored in ionic liquids at room temperature for a longer duration. However, their role is not limited to this and can be expanded to analytical detection methods. In a study by Shi et al. (2012), ionic liquid such as bicyclic 1-butyl-2,3-tetramethyleneimidazolium bromide was found to be more effective than commonly used reagents such as DMSO and betaine for the enhancement of PCR.

# References

Fuchs-Telka, S., Fister, S., Mester, P. J., Wagner, M., & Rossmanith, P. (2016). Hydrophobic ionic liquids for quantitative bacterial cell lysis with subsequent DNA quantification. *Analytical and Bioanalytical Chemistry, 409*(6), 1503–1511. https://doi.org/10.1007/s00216-016-0112-x

Gonzalez García, E., Ressmann, A. K., Gaertner, P., Zirbs, R., Mach, R. L., Krska, R., Bica, K., & Brunner, K. (2014). Direct extraction of genomic DNA from maize with aqueous ionic liquid buffer systems for applications in genetically modified organisms analysis. *Analytical and Bioanalytical Chemistry, 406*(30), 7773–7784. https://doi.org/10.1007/s00216-014-8204-y

Martzy, R., Bica-Schröder, K., Pálvölgyi, D. M., Kolm, C., Jakwerth, S., Kirschner, A. K. T., Sommer, R., Krska, R., Mach, R. L., Farnleitner, A. H., & Reischer, G. H. (2019). Simple lysis of bacterial cells for DNA-based diagnostics using hydrophilic ionic liquids. *Scientific Reports, 9*(1). https://doi.org/10.1038/s41598-019-50246-5

Ressmann, A. K., García, E. G., Khlan, D., Gaertner, P., Mach, R. L., Krska, R., Brunner, K., & Bica, K. (2015). Fast and efficient extraction of DNA from meat and meat derived products using aqueous ionic liquid buffer systems. *New Journal of Chemistry, 39*(6), 4994–5002. https://doi.org/10.1039/c5nj00178a

Shi, Y., Liu, Y. L., Lai, P. Y., Tseng, M. C., Tseng, M. J., Li, Y., & Chu, Y. H. (2012). Ionic liquids promote PCR amplification of DNA. *Chemical Communications, 48*(43), 5325. https://doi.org/10.1039/c2cc31740k

Shamshina, J. L. (2019). Chitin in Ionic Liquids: Historical Insights on the Polymer's Dissolution and Isolation. A Review, *Green Chem. 21*, 3974–3993. https://doi.org/10.1039/C9GC01830A

Shukla, S. K., & Mikkola, J. P. (2020). Use of ionic liquids in protein and DNA chemistry. *Frontiers in Chemistry, 8*. https://doi.org/10.3389/fchem.2020.598662

# Lithium Chloride-Based Isolation of RNA

**Abstract** The extraction of RNA and then isolation of mRNA specifically from the cell homogenate or in vitro transcription reactions can be done easily by lithium chloride (LiCl)-based isolation. This method of RNA isolation is more efficient in precipitating and removing out the unincorporated or smaller fragments of nucleotides. The basic principle involves the interaction between ribose sugar of RNA and positive cations (lithium) to allow the folding and precipitation of RNA only, leaving out the DNA and protein in the aqueous solution. Therefore, there is just simple homogenization of the cell/tissue sample in a suitable buffer, followed by its incubation with LiCl at standardized temperature and pH. The final precipitated RNA can be collected after removing out the aqueous (unprecipitated) sample. In spite of the abovementioned advantages, this technique might not be suitable for isolating the lower concentration of RNAs and different types of RNA other than mRNAs, as evident by the different contradictory studies.

**Keywords** RNA · Lithium chloride · In vitro transcription · UV spectrophotometer · mRNA

## 1 Introduction

Lithium chloride-based RNA precipitation is a fast and convenient method for the extraction of nucleic acids. Since RNA fragments get precipitated only (DNA and protein remain in the solution), this method is suitable for extracting full-length mRNAs from in vitro transcription reactions with a little remnant of free nucleotides (Barlow et al., 1963). For the same reason, UV spectroscopy reading during LiCl precipitated RNA's quantification is more accurate than the RNA isolated by any other procedure (Kistner & Matamoros, 2005). This method is also preferred over ethanol or isopropanol precipitation for RNA isolation. However, experiments like ribonuclease protection assay and molecular cloning may require further purification of the isolated DNA and RNA. Moreover, RNAs of small size (<100

A. Gautam, *DNA and RNA Isolation Techniques for Non-Experts*, Techniques in Life Science and Biomedicine for the Non-Expert, https://doi.org/10.1007/978-3-030-94230-4_8

nucleotides) and with a high degree of secondary structure (e.g., tRNA) are not precipitated by LiCl efficiently.

## 2   Basic Principle

The basis of RNA precipitation by LiCl rather than the precipitation of DNA and proteins is not well-understood. But, as the ribose sugar in RNA makes it more reactive than the DNA, the cations ($Li^+$ in this case) can attack the hydroxyl groups in the ribose backbone and lower the activation energy of hydrolysis. In addition, $Li^+$ may also promote RNA folding by diminishing the repulsion between RNA phosphates. Both of these factors can lead to the selective precipitation of RNA at suitable pH and temperature.

## 3   Protocol

### 3.1   Reagents Required

- Tris base (1 M)
- LiCl
- Spin column
- 0.1% DEPC (diethylpyrocarbonate)
- Distilled water
- EDTA (0.5 M)
- HCl

### 3.2   Working Solution

| Solution | Preparation | Storage |
|---|---|---|
| TE buffer (pH 8, 100 mL) | Mix 1 mL of 1 M Tris base (pH 10–11) and 0.2 mL of 0.5 M EDTA in 98 mL of distilled water. Maintain pH 8 by adding a few drops of concentrated HCl | Room temperature |
| 0.5 M LiCl (10 mL) | Add 0.21 g of LiCl to 10 mL distilled water | Room temperature |
| DEPC water | Mix 1 ml of 0.1% DEPC thoroughly in 1 L distilled water. Autoclave after 1 h and cool | Room temperature |

## 3.3  Procedure

The following methods can be used for the isolation of DNA as well as RNA from the biological samples:

1. Solubilize 1–5 mL of the nucleic acid-containing sample with 0.1 mL of TE buffer in a polypropylene tube.
2. Add 0.4 mL of 0.5 M LiCl to this solution and mix well.
3. Keep the tube at −20 °C for about 30 min for the precipitation.
4. Centrifuge the contents at 4 °C for 20 min at 16,000×$g$.
5. For DNA extraction, collect the supernatant in a separate tube. This DNA can be purified through the Sepharose CL-6B column (equilibrated with TE buffer) and 0.5 M LiCl.
6. The pellet will contain RNA. Air-dry the pellet and resuspend in DEPC-treated distilled water for further use or storage.

# 4  Precautions

The following general precautions may be taken during the procedure:

- Wear gloves during the whole experiment.
- Balance the rotor tubes during centrifugation.
- Collect the supernatant carefully to avoid the contamination of precipitated RNA.
- Handle RNA carefully, as it can be degraded easily by RNase contamination.
- Handle the microbial culture and sample carefully to avoid any surface contamination.

# 5  Applications

- The precipitated RNA can be used for various molecular biology studies, including hybridization experiments and in vitro translation reactions.
- The purified DNA can be used for molecular cloning and DNA sequencing.
- Since this method eliminates free NTPs efficiently, this permits for more accurate quantitation of RNA by UV spectrophotometry.
- It is preferred to eliminate translation inhibitors or prevent cDNA synthesis from RNA preparations (Cathala et al., 1983).
- It can also be used for the enrichment of RNA fragments isolated from other methods.

# References

Barlow, J. J., Mathias, A. P., Williamson, R., & Gammack, D. B. (1963). A simple method for the quantitative isolation of undegraded high molecular weight ribonucleic acid. *Biochemical and Biophysical Research Communications, 13*(1), 61–66.

Cathala, G., Savouret, J. F., Mendez, B., West, B. L., Karin, M., Martial, J. A., & Baxter, J. D. (1983). A method for isolation of intact, translationally active ribonucleic acid. *DNA, 2*(4), 329–335.

Kistner, C., & Matamoros, M. (2005). RNA isolation using phase extraction and LiCl precipitation. In *Lotus japonicus handbook* (pp. 123–124). Springer.

# Isolation of Bacteriophage DNA by PEG Method

**Abstract** The polyethylene glycol (PEG) method is one of the most appropriate approaches for extracting DNA from bacteriophages. The standard method of preparing DNA entails numerous expensive, time-consuming processes and delivers low DNA yields. This procedure, on the other hand, is quite simple and rapid. The PEG method is based on the solubilization of phase particles and macromolecular crowding of DNA, which may then be purified with phenol and precipitated with ethanol. The concentration of PEG required for precipitation is inversely proportional to the size of the DNA fragments. This method can also be used to extract DNA from bacteria and fungi.

**Keywords** Polyethylene glycol · Genetic engineering · Vaccine · Bacteriophage · *E. coli*

## 1 Introduction

Bacteriophages (or phages) are known as bacteria-eating viruses because they infect and destroy the viruses. Phages are composed of proteins that encapsulate a DNA or RNA genome (Fig. 1). They are regarded as potent antibacterial agents due to their ease of availability, natural existence, specificity in their activity, and ability to multiply rapidly in the presence of their host. Due to the capability of transferring their DNA into the host cell genome (transduction), phage DNA is widely used as a vector for genetic engineering. These days, they are increasingly being used in biological and molecular applications like medicine, food engineering, and biotechnology. Therefore, there is a constant need for the isolation of phage DNA with suitable procedures. The traditional process of DNA preparation, which involves CsCl (cesium chloride) gradient purification of the virus, subsequent serial phenol extraction, and ethanol precipitation, is costly, is time-consuming, and generates poor yields of DNA. PEG (polyethylene glycol) is the method for quick isolation of

© The Author(s), under exclusive license to Springer Nature Switzerland AG 2022       73
A. Gautam, *DNA and RNA Isolation Techniques for Non-Experts*, Techniques in
Life Science and Biomedicine for the Non-Expert,
https://doi.org/10.1007/978-3-030-94230-4_9

**Fig. 1** A general outline
of a bacteriophage particle.
7USSR7, CC BY 4.0
https://creativecommons.
org/licenses/by/4.0, via
Wikimedia Commons.
(Adapted from Rossmann
et al., 2005)

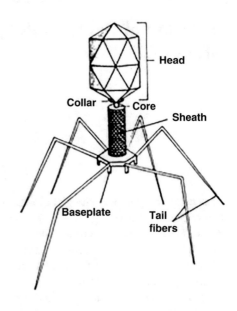

high-quality DNA (Green & Sambrook, 2017; Lockett, 1990). This method is simple, virtually unlimited, and inexpensive (Paithankar & Prasad, 1991).

## 2   Basic Principle

PEG is a straight-chain polymer of a simple repeating unit, $H(OCH_2CH_2)n$ OH. It is a
water-soluble synthetic polymer that is nontoxic and widely utilized in the chemical
and biomedical sectors. It causes changes in DNA polymers' conformation, promot-
ing the fusion and hybridization of the cell (Atha & Ingham, 1981). PEG causes macromolecular crowding of solutes in an aqueous solution; therefore, it is frequently
employed in molecular cloning. The concentration of PEG required for precipitation
is inversely proportional to the size of the DNA fragments (Lis & Schleif, 1975).

## 3   Protocol

### 3.1   Reagents Required

- Sodium acetate (3 M, pH 5.2)
- TE buffer 10X (pH 8)
- Cultures of *E. coli* infected with bacteriophage

- Equilibrated phenol
- Ethanol (70%)
- Polyethylene glycol 8000 (20% w/v) in 2.5 M NaCl

## 3.2 Working Solutions

| Solution | Preparation | Storage |
| --- | --- | --- |
| TE buffer (pH: 8)<br>– Tris: 10 mM<br>– EDTA: 1 mM | 1 mL from 1 M Tris-HCl stock +0.2 mL of 0.5 M EDTA stock and make the final volume to 100 mL | Room temp. |
| Polyethylene glycol 8000 (20% w/v) in 2.5 M NaCl | Mix 25 mL of 40% PEG 8000 in 25 mL of 5 M NaCl to make volume to 50 mL. It will result in a final concentration of 20% PEG in 2.5 M NaCl | 4 °C |
| 3 M sodium acetate (1 L) | In 500 mL of deionized $H_2O$, dissolve 246.1 g sodium acetate. Adjust the pH to 5.2 using glacial acetic acid. Cool the solution overnight; then use glacial acetic acid again to adjust the pH to 5.2. Make up the final volume to 1 L with deionized water. Filter and sterilize the solution before storage | 4 °C |
| Equilibrated phenol | Use as purchased; no preparation needed | Room temp. |
| Ethanol (70%) | Add 737 mL of 95% ethanol (ethyl alcohol), and then add deionized water to bring the total volume to 1000 mL | 4 °C |

## 3.3 Procedure

### 3.3.1 Precipitation of Bacteriophage Particle with PEG

1. Transfer 1 mL of infected log-phase *E. coli* culture in tube and centrifuge at 4000×*g* for 5 min at room temperature.
2. Transfer supernatant to fresh microcentrifuge tube at room temperature.

   Note: *0.1 mL of the supernatant can be stored as the master stock of the bacteriophage for a month.*

3. Add 200 μL of 20% PEG in 2.5 M NaCl to the supernatant.
4. Invert the tube to mix the solution, and then let it stand at room temperature for 15 min.
5. Centrifuge at 8000×*g* for 5 min at 4 °C in a microcentrifuge to recover the precipitated bacteriophage particles.
6. Using a disposable pipette tip/Pasteur pipette, carefully remove all supernatant.
7. Remove the leftover supernatant with a brief 30-s spin.

### 3.3.2   Extraction of Phage DNA with Phenol

1. Resuspend the pellet in 100 μL TE buffer at pH 8.0 by vortexing for 15–30 min at room temperature.
2. Add 100 μL of equilibrated phenol mixture and vortex for 30 s, then place the sample in a standing position for 1 min, and then vortex for another 30 s.
3. Centrifuge the sample for 3–5 min at room temperature at 10,000×g in a microcentrifuge.
4. Transfer the upper aqueous phase to a fresh microcentrifuge tube.

### 3.3.3   Precipitation of Phage DNA

1. Bacteriophage DNA may be precipitated by the standard method using 70% ethanol in the presence of 0.3 M sodium acetate.

   Note: *The aqueous phase may be cloudy when transferred to the fresh micro-centrifuge tube. It should be cleared once sodium acetate solution is added.*

2. Vortex the solution and incubate for 15–30 min at ambient temperature or overnight at −20 °C.
3. Precipitate the bacteriophage DNA by centrifugation at 10,000×g for 10 min at 4 °C.
4. Gently remove the supernatant without disturbing the DNA pellet, which appears as a haze on the tube's side.
5. Centrifuge briefly for 15 s to remove any leftover supernatant.
6. Add 200 μL of chilled 70% ethanol to the pellets.
7. Allow leftover ethanol to evaporate by centrifugation at 10,000×g for 5–10 min.
8. Dissolve DNA pellet in 40 μL of TE buffer.
9. Heat the solution for 5 min at 37 °C and store the DNA solution at −20 °C.

## 4   Precautions

- The infected bacterial cultures must be incubated in a rotary shaker incubator for 4–6 h at 37 °C.
- Ensure that the whole thick PEG/NaCl solution is thoroughly mixed with the infected cell media.
- Ensure that all traces of supernatant have been extracted from the microcentrifuge tube.
- It is critical to work swiftly and carefully to prevent loss of DNA because the pellet is not firmly adhered to the tube wall after ethanol precipitation.
- Make sure to sterilize solutions by autoclaving for 20 min at 15 psi (1.05 kg/cm²).

- Store buffers at room temperature after covering them with aluminum foil.
- It is critical to thoroughly resuspend the bacteriophage pellet to efficiently extraction of DNA using phenol.

# 5   Applications

## 5.1   Genetic Engineering

One of the most common applications of bacteriophage DNA extraction is gene cloning. Bacteriophages have been identified as a special type of nano vehicle for the delivery of genes because they possess a unique DNA packaging system. Bacteriophage particles can be used to target therapeutic agents and the genetic sequences as drug delivery systems. Bacteriophages are considered a suitable cloning vector or gene carrier; their specific vector DNA transfers a fragment of foreign DNA into a suitable host.

## 5.2   Molecular Studies

DNA extraction followed by gene cloning is also helpful in many molecular studies. One of the frequently used molecular techniques is phage display, which creates polypeptides with unique properties. The polypeptide-encoding DNA is fused to the genes of phage coat protein, and the desired protein is expressed on the phage particle's surface. As reagents for investigating molecular recognition, these peptides can be used in drug design. By inhibiting receptor-ligand binding or acting as an agonist, they can also be used as therapeutic drugs (Sidhu, 2000). The filamentous *E. coli* phage M13 is commonly used for phage display. Other phages, such as T7 and lambda, are also employed in phage display systems.

## 5.3   Vaccine Development

Vaccines have been delivered using phages as vehicles. The vaccination antigens expressed on the surfaces of phage particles can be utilized directly. However in DNA vaccines, vaccine antigen-producing sequences are incorporated into phage DNA, and the phage is used as a vehicle to administer DNA vaccines (Clark & March, 2004). The phage display technique can be utilized to create phages with antigenic peptides expressed on their surfaces.

# References

Atha, D., & Ingham, K. (1981). Mechanism of precipitation of proteins by polyethylene glycols. Analysis in terms of excluded volume. *Journal of Biological Chemistry, 256*(23), 12108–12117. https://doi.org/10.1016/s0021-9258(18)43240-1

Clark, J. R., & March, J. B. (2004). Bacterial viruses as human vaccines? *Expert Review of Vaccines, 3*(4), 463–476. https://doi.org/10.1586/14760584.3.4.463

Green, M. R., & Sambrook, J. (2017). Preparation of single-stranded bacteriophage M13 DNA by precipitation with polyethylene glycol. *Cold Spring Harbor Protocols, 2017*(11). https://doi. org/10.1101/pdb.prot093419

Lis, J. T., & Schleif, R. (1975). Size fractionation of double-stranded DNA by precipitation with polyethylene glycol. *Nucleic Acids Research, 2*(3), 383–390. https://doi.org/10.1093/ nar/2.3.383

Lockett, T. J. (1990). A bacteriophage λ DNA purification procedure suitable for the analysis of DNA from either large or multiple small lysates. *Analytical Biochemistry, 185*(2), 230–234. https://doi.org/10.1016/0003-2697(90)90284-g

Paithankar, K., & Prasad, K. (1991). Precipitation of DNA by polyethylene glycol and ethanol. *Nucleic Acids Research, 19*(6), 1346. https://doi.org/10.1093/nar/19.6.1346

Rossmann, M. G., Morais, M. C., Leiman, P. G., & Zhang, W. (2005). Combining X-ray crystallography and electron microscopy. *Structure, 13*(3), 355–362. https://doi.org/10.1016/j. str.2005.01.005

Sidhu, S. S. (2000). Phage display in pharmaceutical biotechnology. *Current Opinion in Biotechnology, 11*(6), 610–616. https://doi.org/10.1016/s0958-1669(00)00152-x

# DNA Isolation by Chelex Method

**Abstract** The Chelex method of DNA extraction is suitable for extracting the DNA from a smaller amount of samples. This method is quick and straightforward and does not involve any harmful organic solvents. The basic protocol involves the extraction of DNA by adding samples to hot Chelex suspensions at pH 10–11. The alkalinity of resin suspension and exposure to heat result in disruption of the cell membrane. Heating also causes the double helix of DNA to denature. Resin beads bind to the cellular components, while DNA (and RNA) remains dissolved in the aqueous solution. Chelex resin also inhibits DNA degradation by chelating metal ions. This method is particularly beneficial for forensic applications, but it is not appropriate for large-scale DNA extraction. The resulting single-stranded DNA is less stable, therefore, not suitable for long-term storage and RFLP analysis.

**Keywords** Chelex 100 · Forensic science · Pathology · Chelating resin · PCR

## 1 Introduction

The Chelex technique of DNA isolation is simple, quick, and free of hazardous chemical solvents. It extracts DNA more efficiently than proteinase K, phenol-chloroform, and silica-based methods. Although previously described organic and solid-phase extraction methods such as phenol-chloroform and silica-based techniques successfully extract high molecular weight DNA from larger samples, these involve several steps of transfer in additional containers for washing or desalting purpose. One of the simplest and least costly techniques for extracting DNA is to use Chelex 100 resin because alternative approaches need multiple steps of transfer in several containers to eliminate contaminants; it gives researchers more control over the experiments and simplifies troubleshooting. An additional step in the Chelex method is proposed by Singh et al. (2018), providing more pure DNA for PCR reaction. This technique is beneficial in forensic science and pathology, such as identifying bacterial or viral DNA in human blood.

© The Author(s), under exclusive license to Springer Nature Switzerland AG 2022
A. Gautam, *DNA and RNA Isolation Techniques for Non-Experts*, Techniques in
Life Science and Biomedicine for the Non-Expert,
https://doi.org/10.1007/978-3-030-94230-4_10

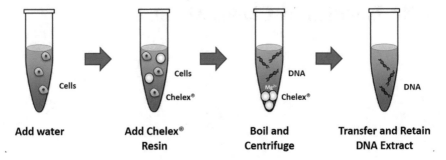

**Fig. 1** Steps in Chelex method of DNA extraction. (Adapted from McKiernan & Danielson, 2017)

## 1.1 *Basic Principle*

Walsh et al. (2013) first established the technique for DNA extraction using Chelex 100. It has been previously proposed to rapidly extract human DNA for polymerase chain reaction (de Lamballerie et al., 1992). It is a chelating resin, having a high affinity for polyvalent metal ions. It is made up of styrene-divinylbenzene copolymers with paired iminodiacetate ions (acts as chelating groups). When cells are heat denatured in a Chelex 100 resin solution, high temperature helps DNA be released into the solution and makes it easier for Chelex resin to attach to it. In addition, Chelex inhibits DNA degradation by chelating metal ions that function as catalysts in the breakdown of DNA (Singer-Sam et al., 1989) (Fig. 1).

Two additional steps are used to purify and concentrate the DNA (Singh et al., 2018). In these steps, ammonium acetate is used for protein precipitation and sodium acetate isopropanol mixture for DNA precipitation.

## 2 Protocol

This protocol is adapted from Singh et al. (2018).

## 2.1 *Reagents Required*

- 0.5 M EDTA (pH 8.0)
- 10X TE (Tris EDTA) buffer (pH 8.0)
- 5% Chelex suspension (pH 8.0, to be prepared in 1X TE buffer)
- 7.5 M ammonium acetate
- 3 M sodium acetate

- 10 N NaOH (make in autoclaved double distilled water)
- 70% ethanol
- Isopropanol

## 2.2   Working Solutions

| Solution | Preparation | Storage |
|---|---|---|
| 0.5 M EDTA | Add 186.1 g of EDTA to 800 mL of distilled water, stir vigorously on a magnetic stirrer and adjust pH to 8 by using NaOH | Room temperature |
| 5% Chelex suspension | Add 4 g of Chelex resin in 40 mL TE buffer and add 40 mL of sterilized water | 4 °C |
| 10X TE buffer (pH 8) | Add 200 μL EDTA into 1 mL of tris buffer and add 8.8 mL of autoclaved Milli-Q water | RT |
| 7.5 M ammonium acetate | Use as purchased; no preparation needed | 4 °C |
| 3 M sodium acetate | Use as purchased; no preparation needed | 4 °C |
| 70% ethanol | Add 737 mL of 95% ethanol, and then add deionized water to bring the total volume to 1 L | 4 °C |
| 10 N NaOH | Add 40 g of NaOH pellets to 80 mL of water slowly, with a continuous stir. Maintain final volume to 100 mL by adding more water | RT |

## 2.3   Procedure

### 2.3.1   Preparation of Blood and Tissue Sample

1. Adsorb approximately 5 μL of blood onto a disc of Whatman filter paper (4 mm radius, 180 μm thicknesses).

   Note: *10–15 mg plant or animal tissue should be homogenized with a pestle for 1 min in 300 μL of 5% Chelex 100 in case of performing DNA extraction in animal and plant tissues.*

2. Dry the disc in a laminar hood for 30 min.
3. Place it in an Eppendorf tube for further processing.

### 2.3.2   DNA Isolation Protocol

1. Add 200 μL of 5% stock Chelex solution into a 1.5 mL Eppendorf tube.
2. Heat the tube at 100 °C for 10 min in a boiling water bath.

3. Add blood-soaked filter paper disc or the animal/plant tissue sample to the hot Chelex solution.
4. Heat the tube for 8 min, vortex, and heat again for 7 min.
5. Centrifuge the tube at 12,000×g for 90 s at room temperature.
6. Gently pipette out the supernatant to avoid extracting the Chelex resin.
7. The following steps are required if further purification of DNA is required.

### 2.3.3   Protein Precipitation

1. Add 7.5 M stock solution of ammonium acetate to the supernatant to make the final concentration 2.5 M.
2. A thick yellowish-white precipitate of protein will appear.
3. Allow it to rest for 5 min on ice.
4. Vortex the sample for 5 s and centrifuge at 12,000×g for 10 min at room temperature to pellet out protein.
5. Collect the clear supernatant containing genomic DNA.

### 2.3.4   DNA Precipitation

1. Add 3 M sodium acetate to the supernatant to make the final concentration 0.3 M.
2. Add 200 μL of 75% ice-cold ethanol into the supernatant.
3. Vortex solution for 5 s and allow to stand for 4 h at −30 °C to precipitate DNA.
4. Discard the supernatant and wash the pellet twice with 75% ice-cold ethanol.
5. Centrifuge the tube after each wash at 15,000×g for 10 min at 4 °C.
6. Finally, wash the pellet with 100% ice-cold isopropanol, remove the supernatant, and dry it in a laminar flow hood.
7. After 10 min, add 10 μL of deionized water to the pellet, and incubate at 55 °C for 10 min to facilitate solubilization.
8. Measure the quantity and purity of DNA in each sample with the help of a spectrophotometer and agarose gel electrophoresis.
9. DNA solution can be stored at −20 °C until use.

## 3   Precautions

1. While pipetting Chelex solutions, make sure the resin beads are evenly dispersed in the solution by gently mixing with a stir bar in a beaker.
2. The DNA extracted by the Chelex method is generally contaminated at the end of the procedure, so its purification is necessary.
3. 5% Chelex solution should be adequately stirred so that pellets equally distribute in solution because resin beads are sticky.

4. Ammonium acetate stock solution should be kept at 4 °C and utilized within a month.
5. Prepare 10 N NaOH with extreme care since it includes a very exothermic reaction that might break the glass container.

# 4  Applications

## 4.1  Forensic Science

The Chelex technique of DNA extraction is beneficial for extracting DNA from extremely small tissue samples, dried blood spots, or swabs. As a result, in forensic sciences, this method is favored. Although the DNA samples collected are not particularly pure, they are acceptable for most downstream applications such as PCR, qPCR, and sequencing. When ultrapure DNA is desired, further cleansing techniques may be necessary.

## 4.2  Pathology

Medical pathology is a branch of medicine that deals with the diagnosis and treatment of diseases. Frequently, just a small amount of samples are available. The Chelex technique comes in handy in these situations. The typical technique for detecting many pathogenic organisms in the blood is isolating pathogen DNA and amplifying it. Furthermore, as compared to other approaches, this method is very inexpensive and straightforward. Thus, a large number of samples may be tested in less time.

## 4.3  Gene Sequencing

This approach is used to identify the nucleotide sequence of DNA since nucleotides reflect the essential characteristics of a gene or genome. It is, in fact, a blueprint that contains all of the instructions for creating creatures. DNA extracted using this isolation procedure may be separated by size using electrophoresis and sequenced by an automated sequencing machine where DNA can be labeled with fluorescent dyes. As a result, the appropriate DNA fragment can be obtained and employed in cloning procedures later.

# References

de Lamballerie, X., Zandotti, C., Vignoli, C., Bollet, C., & de Micco, P. (1992). A one-step microbial DNA extraction method using "Chelex 100" suitable for gene amplification. *Research in Microbiology, 143*(8), 785–790. https://doi.org/10.1016/0923-2508(92)90107-y

McKiernan, H., & Danielson, P. (2017). Molecular diagnostic applications in forensic science. *Molecular Diagnostics*, 371–394. https://doi.org/10.1016/b978-0-12-802971-8.00021-3

Singer-Sam, J., Tanguay, R. L., & Rjggs, A. O. (1989). Use of Chelex to improve PCR signal from a small number of cells. *Amplifications: A Forum for PCR Users, 3*(September):11.

Singh, U. A., Kumari, M., & Iyengar, S. (2018). Method for improving the quality of genomic DNA obtained from minute quantities of tissue and blood samples using Chelex 100 resin. *Biological Procedures Online, 20*(1). https://doi.org/10.1186/s12575-018-0077-6

Walsh, P. S., Metzger, D. A., & Higuchi, R. (2013). Chelex 100 as a medium for simple extraction of DNA for PCR-based typing from forensic material. *BioTechniques, 54*(3). https://doi.org/10.2144/000114018

# DNA Isolation by Lysozyme and Proteinase K

**Abstract**  The isolation of DNA using lysozyme and Proteinase K is suitable for extraction of DNA from microorganisms particularly Gram-positive bacteria. The application of enzymes like lysozyme and Proteinase K at suitable temperatures, duration, and pH helps in the degradation of hard and rigid peptidoglycan cell walls of these microorganisms. Therefore, it is the most efficient and quick method for the isolation of DNA from microorganisms without using harsh biochemicals. However, further purification of DNA can be required by other methods for utilizing it further for molecular studies. This method is widely used in microbiology research, production of recombinant DNAs, biotechnology-based industries, and medical diagnosis where the presence of microbial DNA confirms the pathogenicity of a particular disease.

**Keywords**  DNA · Microorganisms · Gram-positive bacteria · Proteinase K · Lysozyme

## 1  Introduction of the Technique

The previous chapters mention that one can isolate nucleic acids from different cells, including microorganisms using guanidinium thiocyanate and phenol-chloroform methods. However, rigid cell walls in some microorganisms like Gram-positive bacteria, yeast, spores, etc. make these DNA extraction methods difficult and inefficient (Vingataramin & Frost, 2015). Moreover, these generally used procedures are tedious and unsuitable for application to large sample numbers. For example, a vast volume of culture cells is required to isolate enough genomic DNA for DNA recombination technology. These limitations can be overcome by the pretreatment of microorganisms with lysozymes and Proteinase-K enzymes for DNA isolation. Lysozymes digest the cell wall components, whereas Proteinase K helps in the elimination of interfering proteins. This method saves the isolated DNA from probable damage, otherwise which can occur in lengthy protocols by repeated use of harsh biochemicals. More purification of the isolated DNA can be done as per its requirement for further experiments/study by suitable precipitation methods like phenol-chloroform-ethanol or column-based protocols (Martzy et al., 2019). DNA from WBCs, soft tissues like muscle, brain, flower, hair follicles, sperm cells, epithelial cells, and osteocytes of bones can also be isolated through this method.


## 2    Basic Principle

The cell walls of Gram-positive bacteria are majorly made up of peptidoglycan, which is present around the cell membrane in multilayered form. The presence of teichoic acid and N-acetylglucosamine cross-links in peptidoglycan layers provides rigidity to their cell walls. Due to this rigidity, the general method of phenol-chloroform-based DNA or RNA isolation is unsuitable for Gram-positive bacteria and other similar microbes. On the other hand, lysozyme is a naturally occurring enzyme that kills bacteria by attacking and damaging the peptidoglycan layer. Proteinase K is a wide-spectrum protease that can digest many surface proteins, including nucleases. The addition of lysozyme breaks opens the cell wall for easier removal of the nucleic acids from the cell along with the suitable antibiotic and/or cell lysis buffer. Further, the addition of Proteinase K aids in the deactivation of RNase/DNase and avoids the degradation of nucleic acids in the solution. In addition, Proteinase K helps in stopping the action of lysozyme in this protocol. Finally, the extracted nucleic acid can be purified either by a suitable column or phenol-chloroform-ethanol protocol.


## 3    Protocol

### 3.1    Reagents Required

- **Tris-HCl 1 M (pH 8)** (store at 4 °C)
- **EDTA 0.5 M solution, pH 8.0** (to be stored at room temperature)
- **Sodium chloride, 5 M solution** (to be stored at room temperature)
- **Lauryl sulfate, 10% solution.** (10% SDS, 10% sodium dodecyl sulfate) (to be stored at room temperature)
- **Tris-EDTA Buffer 100X Concentrate** (1 M Tris-HCl, 0.1 M EDTA, pH 8.0) (to be stored at room temperature)
- **Ribonuclease A**, free of DNase activity (to be stored at 2–8 °C)
- **Proteinase K** (available in lyophilized form)

## 3.2   Working Solutions

| Solution | Preparation | Storage |
|---|---|---|
| Tris-HCl (pH 8) (10 mM), 100 mL | Take 1 mL of 1 M Tris-HCl (pH 8.0) and add 99 mL of deionized water | 4 °C |
| Cell Lysis Buffer (10 mM Tris-HCl, 26 mM EDTA, 17.3 mM (0.5%) SDS), 1000 mL | Add 10 mL Tris-EDTA Buffer, 50 mL EDTA 0.5 M Solution, and 50 mL lauryl sulfate in 890 mL deionized water (pH 7.3) | Room temperature |
| RNase A Solution (4 mg/mL) | Dissolve 10 mg RNAase in 2.5 mL TE buffer | −20 °C |
| Proteinase K | 20 mg/mL in Tris-HCl (pH 8). Thaw before use | −20 °C |
| Lysozyme | Dissolve 25 mg lysozyme powder in 1 mL of 10 mM Tris-Cl, pH 8.0 at 37 °C. Thaw before use | −20 °C |

## 3.3   Protocol for DNA Isolation

The following protocol is a summarized version of the basic protocol (Bollet et al., 1991) and subsequent adaptations (Sohail, 1998; Sambrook and Russell, 2001):

1. Culture the microorganisms in a suitable medium, and take out the growing cells/log-phase cells along with media (~1 to 10 mL) in a 10–50 mL centrifuge tube.
2. Centrifuge the media for 1 min at $12,000 \times g$ at 4 ° C to obtain the microbial cells as a pellet. Discard the supernatant.
3. Resuspend the pellet in 80 μL of 10 mM Tris-Cl (pH 8) by vortexing.
4. Mix 20 μL lysozyme solution in the cell suspension thoroughly.
5. Add 100 μL of cell lysis buffer and vortex the whole solution.
6. Make this solution clear by keeping it at 37 °C for ~10 min.
7. Stop the activity of lysozyme by adding 10 μL Proteinase K and mixing it thoroughly.
8. Incubate the tube in a shaking water bath maintained at 56 °C for half an hour.
9. Finally, to obtain the RNA-free DNA, add 3 μL of RNase A (stock 10 mg/mL, working concentration 100 μg/mL) to the solution, and incubate for 5 min at 56 °C in the same water bath.
10. In some cases, the phenol-chloroform-isopropanol protocol can be followed after RNase A treatment to obtain the precipitated and purified DNA.

## 4   Precautions

- The tubes must be properly balanced during centrifugation.
- It is advisable to wear a lab coat and gloves while handling the biochemicals.

- All the processes which require phenol should be executed in the fume hood.
- The laboratory area must be thoroughly cleaned with 70% alcohol throughout all the procedures of the experiment.
- Store and thaw the enzymes properly to maintain their enzymatic activity intact.
- Handle the microbial culture and sample carefully to avoid any surface contamination.

# 5   Applications

**Molecular epidemiological studies**: The study and analysis of various determinants and frequency patterns of disease by techniques like Southern hybridization, restriction endonuclease digestion, RAPD-PCR, etc. requires the extraction of good-quality DNA from large numbers of bacteria or other samples. In that case, this protocol is helpful for the isolation of pure DNA.

**Molecular biology studies**: Molecular biology studies which require pure and large amounts of DNA like DNA analysis with endonuclease digestion, Southern hybridization, restriction digestion, drug-DNA interaction, PCR, and PCR-based methods can rely on this protocol for DNA from the microorganisms.

# References

Bollet, C., Gevaudan, M. J., De Lamballerie, X., Zandotti, C., & De Micco, P. (1991). A simple method for the isolation of chromosomal DNA from gram positive or acid-fast bacteria. *Nucleic Acids Research, 19*(8), 1955.
Martzy, R., Bica-Schröder, K., Pálvölgyi, Á. M., Kolm, C., Jakwerth, S., Kirschner, A. K., … Reischer, G. H. (2019). Simple lysis of bacterial cells for DNA-based diagnostics using hydrophilic ionic liquids. *Scientific Reports, 9*(1), 1–10.
Sambrook, J., Russell, D. W., & Cold Spring Harbor Laboratory. (2001). *Molecular cloning: A laboratory manual*. Cold Spring Harbor Laboratory.
Sohail, M. (1998). A simple and rapid method for preparing genomic DNA from gram-positive bacteria. *Molecular Biotechnology, 10*(2), 191–193.
Vingataramin, L., & Frost, E. H. (2015). A single protocol for extraction of gDNA from bacteria and yeast. *BioTechniques, 58*(3), 120–125.

# Isolation of DNA from Blood Samples by Salting Method

**Abstract** Extraction of blood genomic DNA is a systematic method in clinical and molecular biology research. Despite the fact that several procedures for extracting blood genomic DNA have been developed, the bulk of these processes are time-consuming and involve expensive chemicals as well as dangerous organic solvents. The majority of the materials used in standard DNA extraction are hazardous, and it is desirable to adopt a successful DNA extraction process that does not need such chemicals. For genetic research, high-quality and high-quantity DNA extraction is essential. It is critical to address the usage of easy and low-cost DNA extraction technologies in large-scale gene polymorphism investigations. This chapter aims to examine the most successful way of extracting DNA using the salt-out method. This method is easy, quick, safe, and inexpensive. It may be utilized in medical laboratories and research facilities.

**Keywords** DNA isolation · Salting-out method · Cell lysis · Genomic DNA

## 1 Introduction

As mentioned in the earlier chapters, the cells from any sample are separated from each other often by a physical means such as grinding or vortexing and put into a solution containing salt. The salting-out method is based on separating DNA from cells using a solution containing salt (Maurya et al., 2013). This superior and inexpensive extraction method is generally used for DNA extraction from the peripheral blood through the salting-out purification technique (Miller et al., 1988).

## 2   Basic Principle

The salting-out method is mainly used for the precipitation of large biomolecules such as protein. Isopropanol precipitation is commonly used for concentrating and desalting nucleic acid (DNA or RNA) preparations in an aqueous solution (Nasiri et al., 2005). The following are the basic steps involved in the isolation of DNA:

- **Separation of cells from blood samples**: Cells from which the DNA is extracted are separated from the rest of the unwanted cells.
- **Cell lysis**: The cell membranes and nuclear membrane are lysed by the action of buffer and detergents.
- **Precipitation**: DNA is precipitated/extracted from the rest of the solution.
- **Purification/washing**: The precipitated DNA is washed using alcohol.

## 3   Protocol

For the DNA extraction from the blood sample, a modified protocol of Suguna et al. (2014) has been mentioned below.

### 3.1   Reagents Required

- Blood samples should be collected in a vacutainer containing ethylenediamine-tetraacetic acid (EDTA).
- TKM (Tris-KCl-MgCl$_2$) and buffer containing Tris-HCl.
- Potassium chloride.
- Magnesium chloride.
- EDTA.
- Triton-X 1%.
- Sodium chloride.
- Sodium dodecyl sulfate (SDS).
- Isopropanol.
- Ethanol.
- TE buffer.

### 3.2   Working Solutions

| Solution | Preparation | Storage |
|---|---|---|
| TKM 1 buffer/ low salt buffer (500 mL) | 0.605 g of Tris-HCl (10 mM) pH 7.6, 0.372 g of KCl (10 mM), 1.016 g of MgCl$_2$ (10 mM), 0.372 g of EDTA (2 mM) were dissolved in 500 mL of distilled water | At room temperature (RT) |

| Solution | Preparation | Storage |
|---|---|---|
| TKM 2 buffer/ high salt buffer (100 mL) | 0.121 g of Tris-HCl (10 mM) pH 7.6, 0.074 g of KCl (10 mM), 1.203 g of MgCl$_2$ (10 mM), 0.074 g EDTA (2 mM), 0.467 g of NaCl (0.4 M) were dissolved in 100 mL of distilled water | RT |
| Triton-X (10 mL) | Added 0.1 mL of 100% Triton-X to 9.9 mL of distilled water | RT |
| SDS (10%) | 1 gm of sodium dodecyl sulfate was dissolved in 10 mL distilled water | RT |
| 6 M NaCl | 8.765 g of NaCl was dissolved in 25 mL of distilled water | RT |
| TE buffer | 0.030 g of Tris-HCl (10 mM) pH 8.0, 0.009 g of EDTA (1 mM) were dissolved in 100 mL of distilled water | RT |

## 3.3 Procedure for DNA Extraction

### 3.3.1 RBC Lysis

(a) Take 3 mL of TKM1 buffer +1 mL of blood sample + 167 µL of 1X Triton-X in the autoclaved microcentrifuge tube.
(b) Mix all reagents by using a vortex (Fig. 1A).
(c) Incubate the tube at 37 °C for 5 min in the water bath to lyse the RBCs.
(d) Centrifuge (Fig. 1B) at 5000×$g$ for 5 min.
(e) Discard the supernatant using the pipette, and keep the red-colored pellet.
(f) Add 3 mL of TKM1 + 142 µL of 1x Triton-X.
(g) Mix all reagents by vortexing.
(h) Incubate sample at 37 °C for 5 min in a water bath (Fig. 1C) to lyse the remaining RBCs.

**a) Vortexer**       **b) Centrifuge**       **c) Water bath**

**Fig. 1** Various equipment used for blood DNA extraction by salting-out method. Magnus Manske, CC BY 1.0 https://creativecommons.org/licenses/by/1.0, via Wikimedia Commons

(i) Centrifuge at 5000×g for 5 min to obtain WBCs in the pellet.
(j) Discard supernatant by using the pipette.

### 3.3.2 WBC Lysis

(k) To the WBC pellet, add 1 mL TKM2 + 133 μL SDS (10%) and vortex the tube thoroughly to dissolve the pellet.
(l) Incubate at 37 °C for 5 min in a water bath to lyse the WBCs.
(m) At the end of incubation, 335 μL (6 M) NaCl was added to precipitate the protein.
(n) Centrifuge the tube at 5000×g for 10 min.

### 3.3.3 Precipitation of DNA

(o) Transfer the supernatant to a new tube, and add 500 μL chilled isopropanol to precipitate the DNA.
(p) Spooled out the DNA and transferred it into a 1.5 mL microfuge tube.

### 3.3.4 Washing of DNA

(q) Add 500 μL of 70% ethanol and tap and flick to remove any excess salt.
(r) Centrifuge at 9000×g for 3 min, and discard the ethanol.
(s) Add 500 μL of 80% ethanol for rewashing, and discard the ethanol.
(t) Air-dry the microcentrifuge tube containing washed DNA.

### 3.3.5 Storage of DNA

(u) Add 50 μL TE buffer in a 1.5 mL tube for DNA storage.
(v) Store the DNA at 4 °C for a day (for the complete dissolution of DNA in TE buffer).
(w) Then, store in −20 °C or −80 °C for long-term storage.

## 4  Precautions

- Add buffers in the correct order so that the sample is bound, washed, and eluted in the correct sequence.
- Ethanol precipitation and drying should be done carefully.
- Use careful inversion mixing after cell lysis to avoid shearing DNA.

- Give proper washing to ensure complete removal of ethanol.5. The sample should be fresh for better quality and quantity of DNA.
- Buffers should be prepared carefully that should maintain proper pH and salt concentration.
- Pellet should be carefully dissolved properly for elution of DNA/RNA.

# 5  Applications

- Extracted genomic DNA can be used for the diagnosis of bacterial, viral, fungal, and genetic diseases.
- Extracted genomic DNA is suitable for next-generation sequencing.
- It can be used for genetic fingerprinting in medical science.
- Extracted DNA has several downstream applications in various techniques such as polymerase chain reaction (PCR), DNA cloning, electrophoresis, restriction fragment length polymorphism (RFLP), random amplification of polymorphic DNA (RAPD), amplified fragment length polymorphism (AFLP).

# References

Maurya, R., Kumar, B., & Sundar, S. (2013). Evaluation of salt-out method for the isolation of DNA from whole blood: A pathological approach of DNA based diagnosis. *Int J Life Sci Biotechnol Pharma Res, 2*, 53–57.

Miller, S., Dykes, D., & Polesky, H. (1988). A simple salting out procedure for extracting DNA from human nucleated cells. *Nucleic Acids Research, 16*, 1215.

Nasiri, H., Forouzandeh, M., Rasaee, M. J., & Rahbarizadeh, F. (2005). Modified salting-out method: High-yield, high-quality genomic DNA extraction from whole blood using laundry detergent. *Journal of Clinical Laboratory Analysis, 19*(6), 229–232.

Suguna, S., Nandal, D., Kamble, S., Bharatha, A., & Kunkulol, R. (2014). Genomic DNA isolation from human whole blood samples by non enzymatic salting out method. *International Journal of Pharmacy and Pharmaceutical Sciences, 6*, 198–199.

# CTAB or SDS-Based Isolation of Plant's DNA

**Abstract** Unlike the animal cell, DNA extraction from plant cells faces challenges of a rigid cellulose cell wall and the presence of a variety of cytoplasmic components like polysaccharides, polyphenols, lipids, and proteins. This demands modification in the DNA extraction protocol to obtain DNA in the purest form possible. The major modification includes the replacement of SDS with CTAB, which, apart from being a detergent, helps in the precipitation of polysaccharides under high salt concentrations. Additionally, PVP is also used during grinding as it helps in the removal of polyphenols. Both polysaccharides and polyphenols, if not removed effectively, would coprecipitate with DNA. Due to the high viscosity of CTAB, isolation buffer is maintained at 65 °C, and the homogenized tissue is also incubated at the same temperature. Post incubation in isolation buffer, DNA is extracted into the aqueous phase by chloroform-isoamyl alcohol treatment, and then it is precipitated with isopropanol. After washing this precipitated DNA in 70% ethanol, it is air-dried and resuspended in TE buffer for further use. Depending upon the cytoplasmic composition and the purpose of DNA extraction, protocols are modified for best results. Some of these modified protocols have been discussed here.

**Keywords** DNA extraction · Plants · Protocol · CTAB · PVP · Sequencing · High-quality DNA

## 1 Introduction

DNA, the genetic material of all plants, is housed inside the nucleus like any other eukaryote. What makes its isolation interesting is the surrounding cytoplasmic environment and a rigid cellulosic cell wall. Biochemical heterogeneity among plant species gives rise to variable cytoplasmic environments. For example, cereals are rich in carbohydrates, pulses are rich in proteins, oilseeds are rich in lipids, and medicinal plants are rich in polyphenols and secondary metabolites; these interfere during DNA isolation and have to be eradicated. The amount of these contaminants

© The Author(s), under exclusive license to Springer Nature Switzerland AG 2022
A. Gautam, *DNA and RNA Isolation Techniques for Non-Experts*, Techniques in Life Science and Biomedicine for the Non-Expert,
https://doi.org/10.1007/978-3-030-94230-4_13

in the sample depends on the environmental condition, age of the plant, part of the plant used, and the plant species. DNA extracted from woody samples and herbarium specimens need certain other variations in the protocol. Hence, the DNA extraction protocol needs minor variations to obtain DNA in the purest form possible from the varied cytoplasmic environments. Moreover, the protocol variations are determined by the purpose for which DNA is extracted, i.e., cloning and sequencing (demands high purity) or PCR (sufficiently purified DNA works).

Doyle and Doyle first introduced rapid DNA isolation from fresh plant tissue using CTAB in 1987. After that, researchers have introduced minor modifications to cater to the biochemical heterogeneity among plants. The method was reported to work for many different angiosperm groups (both monocots and dicots) and recently dried herbarium specimens (Doyle & Doyle, 1987).

## 2   Basic Principle

Successful DNA extraction involves four essential steps: (a) disruption of tissue, cell wall (by liquid nitrogen and CTAB), cell membrane, and nuclear membrane (by surfactants), (b) release of intact DNA into solution, (c) precipitation of DNA, and (d) purification from contaminants like polysaccharides, proteins, lipids, secondary metabolites, RNA, and salts (used during extraction).

Tris-HCl, EDTA, CTAB/SDS, PVP-40, ME, NaCl, chloroform, isoamyl alcohol, and ethanol are the several chemicals used to isolate pure DNA. Each of these chemicals has a specific role which is discussed here (Jadhav et al., 2015).

*Tris* (hydroxymethyl aminomethane) is used to maintain the pH of the buffer between 7 and 9. It also interacts with lipopolysaccharides in the cell membrane and assists in its lysis. During grinding, the cellular compartmentalization ends, and the cytoplasmic content is released, which alters the pH, which can affect the stability of biomolecules like nucleic acids. Tris plays an essential role in avoiding this.

*EDTA* (ethylenediaminetetraacetic acid) is a hexadentate chelating agent which binds divalent ions like $Mg^{2+}$ and $Ca^{2+}$. $Mg^{2+}$, the cofactor for the DNase enzyme, gets chelated in the presence of EDTA, making the enzyme nonfunctional, thus protecting DNA. Secondly, sequestration of $Mg^{2+}$ prevents the association of proteins with DNA. Similarly, sequestration of $Ca^{2+}$ loosens up middle lamella (cementing material between cell walls) and helps in easy disruption of the tissue.

*CTAB* (cetyltrimethylammonium bromide) is a cationic detergent soluble in water and readily soluble in alcohol. It has a long hydrophobic hydrocarbon chain and a hydrophilic head that forms micelle in water. During DNA extraction, it captures the phospholipids (these too are amphipathic) from biological membranes, thus rupturing them and releasing the DNA. It is also known to denature enzymes by adding a net positive charge, thus clearing all DNases and RNases that might hydrolyze nucleic acids. Secondly, plant cells have large amounts of polysaccharides that coprecipitate with DNA during the extraction process, giving it a viscous glue-like appearance. Under high salt concentration, CTAB binds to these polysaccharides and precipitates them out, whereas DNA remains soluble. Hence, it is a better-preferred detergent during plant DNA extraction.

*SDS* (sodium dodecyl sulfate) is an anionic detergent that denatures the proteins in the biological membranes resulting in their disruption and release of cellular contents. As SDS linearizes the proteins, it also leads to the dissociation of nucleic acid-protein complexes.

*ME* (**β**-mercaptoethanol) is a potent reducing agent, so it is added to the extraction buffer most of the time to remove tannins and polyphenols. They also denature proteins by wreaking the disulfide bonds and disrupting their tertiary and quaternary structure, thus reducing their solubility.

*PVP* (polyvinylpyrrolidone) is added to the extraction buffer to remove phenolic compounds (polyphenols) from cell extracts. PVP complexes with polyphenols are removed after centrifugation with chloroform from the interphase. In the absence of PVP, polyphenols bind DNA irreversibly and get coprecipitated as contaminants. Secondly, polyphenols that get released from vacuole during cell lysis get oxidized by polyphenol oxidase, PPO (generally located in plastids) forming brown-colored quinone. The presence of PVP also prevents sample discoloration.

*NaCl* (sodium chloride)—NaCl leads to the removal of proteins associated with DNA and helps them remain dissolved in the aqueous layer. Simultaneously, $Na^+$ neutralizes DNA by forming an ionic bond with the negatively charged phosphate group resulting in the coming together of the neutral DNA molecules (helps in precipitation), avoiding the formation of protein-nucleic acid complexes.

*Phenol*: Neutral phenol along with chloroform helps in the removal of proteins and polysaccharides from the DNA preparation. In an aqueous medium, the polar residues of protein form a protective shell around the hydrophobic core (nonpolar amino acid stretches). On adding phenol to the cell extract and shaking it vigorously, the proteins get denatured, exposing the hydrophobic core, and get precipitated in the interphase.

*Chloroform* ($CHCl_3$), a nonpolar solvent, dissolves the cell debris, lipids, and nonpolar proteins in it, leaving the isolated DNA in the upper aqueous phase.

*Isoamyl alcohol*: While extracting the aqueous phase containing DNA, chloroform comes in contact with air. It forms a harmful gas called phosgene ($COCl_2$) that appears as froth in the interphase and hinders proper DNA extraction. Isoamyl alcohol stabilizes the interphase between the aqueous and organic (chloroform) phases.

*Ribonuclease A* is an endonuclease that hydrolyses RNA. During DNA purification, this step helps in the removal of RNA contamination. RNase is a heat-stable enzyme, so it is recommended to keep it in a boiling water bath for 10 mins as it helps to get rid of any DNase contamination.

*Isopropanol* and *ethanol* remove the hydration shell around the phosphate, leading to the precipitation of the neutralized DNA (having $Na^+$ bonded to the phosphate group) from the solution. But they cannot break the hydration shell of RNA due to the 2'OH, which is firmly bound to water. As a result, RNA remains soluble in isopropanol, enabling selective precipitation of DNA. Isopropanol also dissolves chloroform, thus removing the impurities of the previous steps.

*Sodium acetate, ammonium acetate, or potassium acetate*: As discussed earlier, Na⁺ neutralizes DNA and helps them come together. If this is made to occur in the presence of isopropanol or ethanol, much better precipitation will be obtained. As acetates of the monovalent cations are more soluble in isopropanol, their use is recommended during precipitation.

*70% ethanol* is used to remove excess salt in the DNA pellet, which had come during the DNA extraction process.

# 3   Protocol

## 3.1   Reagents Required

- Tris-HCl buffer stock/concentrate (1 M Tris-HCl, pH: 8). Store at room temperature (RT).
- EDTA-sodium salt concentrate (0.5 M, pH: 8); store at RT.
- NaCl.
- CTAB.
- ME.
- PVP-40.
- Chloroform-isoamyl alcohol (24:1) at 2–8 °C.
- Isopropanol at −20 °C.
- Ethanol (95%).
- RNase A.
- Phenol-chloroform-isoamyl alcohol (25:24:1).
- Sodium acetate, 3 M, pH: 5.2.
- Ammonium acetate, 7.5 M.

## 3.2   Working Solutions

| Solution | Preparation | Storage |
|---|---|---|
| Isolation/extraction buffer (pH: 8)<br>– Tris-HCl, 100 mM<br>– EDTA, 20 mM<br>– CTAB, 2%<br>– NaCl, 1.4 M<br>– ME, 0.2% (when a large amount of phenolics are present in the sample, add–PVP-40, 1%) | –10 mL from 1 M Tris-HCl stock +4 mL of 0.5 M EDTA stock + CTAB 2 g + 28 mL of 5 M NaCl stock and add deionized water just below 100 mL<br>– Warm the solution to 65 °C with constant stirring to dissolve CTAB<br>– After complete dissolution, make the final volume to 100 mL<br>– Autoclave and store the solution<br>– Add ME add just before use | 2–8 °C |
| NaCl, 5 M | Dissolve 29.2 g NaCl in deionized water; make the final volume to 100 mL and autoclave | 2–8 °C |

| Solution | Preparation | Storage |
|---|---|---|
| 70% ethanol | To 737 mL of 95% ethanol, add deionized water and make the final volume to 1 L | 2–8 °C |
| Resuspension buffer or TE buffer (pH: 8)<br>– Tris: 10 mM<br>– EDTA: 1 mM | 1 mL from 1 M Tris-HCl stock +0.2 mL of 0.5 M EDTA stock and make the final volume to 100 mL | 2–8 °C |
| RNase A (10 mg/mL) | Dissolve 10 mg RNase A in 1 mL TE buffer (pH: 7.5) having 15 mM NaCl. Heat at 100 °C for 15 min to destroy any DNase contamination. Cool to RT | Store as aliquots in −20 °C |
| Phenol-chloroform-isoamyl alcohol (25:24:1), saturated with 10 mM TE buffer, pH: 8 | Use as purchased | 2–8 °C |
| Sodium acetate (3 M) | Use as purchased | 2–8 °C |
| Ammonium acetate, 7.5 M | Use as purchased | 2–8 °C |

## 3.3 Procedure

DNA is widely extracted from plants using CTAB as the detergent to provide good yield and quality of DNA. Though SDS has been used as a detergent in many protocols (Chiong et al., 2017; Xia et al., 2019; Tamari et al., 2013), CTAB is still the preferred choice.

Doyle and Doyle (1987) protocol with minor modifications is as follows:

1. Prechill pestle and mortar at −20 °C.
2. Prepare 10 mL of extraction buffer by finally adding 20 μL ME (and 0.1 g PVP-40, if required) in a 50 mL centrifuge tube and preheat it at 65 °C in a water bath.
3. Grind 1 g leaf in liquid nitrogen and transfer the powdered tissue to a 50 mL centrifuge tube.
4. Add the preheated extraction buffer to the powdered tissue and mix vigorously.
5. Incubate at 65 °C in a water bath for 30 min and mix by gentle inversion intermittently.
6. Allow the sample to come to room temperature, and add an equal volume of chloroform-isoamyl alcohol (24:1). Gently mix by making the figure "8."
7. Centrifuge at 12,000×g for 10 min at room temperature (RT).
8. Transfer the aqueous phase into a fresh tube using a wide-bore pipette (Fig. 1).
9. Repeat steps 4–6 till the aqueous layer becomes clear.
10. Measure the sample volume and add ⅔ volume of chilled isopropanol (−20 °C). Mix gently by inversion and let the DNA precipitate for 30 min at −20 °C (Fig. 2).

**Fig. 1** Microfuge tube after PCI treatment or chloroform-isoamyl alcohol treatment. (Source: Mphhbxw.jpg)

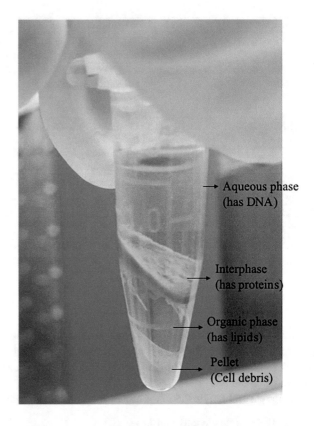

Aqueous phase (has DNA)

Interphase (has proteins)

Organic phase (has lipids)

Pellet (Cell debris)

**Fig. 2** Clumping of DNA strands and their precipitation after adding chilled isopropanol or ethanol. (Source: 5335_dnaComingOut.jpg)

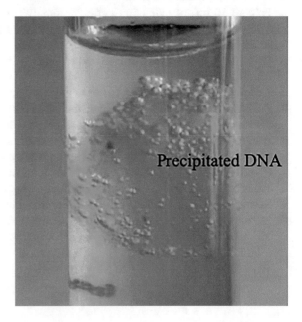

Precipitated DNA

**Fig. 3** Spooling of the precipitated DNA strands with a hook. (Source: 5337_hookDNA.jpg)

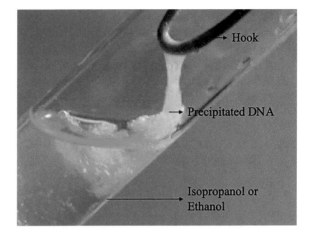

11. If DNA has precipitated nicely, spool it out (Fig. 3) and transfer into a 2 mL microcentrifuge tube, else centrifuge at 10,000 rpm for 10 min at RT and decant the supernatant very carefully.
12. Wash the precipitated DNA in 0.5 mL chilled (−20 °C) 70% ethanol by gently mixing it at RT for 15 min (this washing helps in the removal of residual salts and increases the purity of DNA).
13. Centrifuge at 12,000×g for 10 min at RT and discard the supernatant very carefully.
14. Repeat the ethanol washing once again and vacuum dry or air-dry the pellet containing the precipitated DNA.
15. Resuspend DNA in 1 mL of TE buffer.
16. Add 10 μL RNAse A and incubate in a water bath at 37 °C for 1 hr. (At this stage, DNA is resistant to digestion by endonucleases.)
17. PCI treatment: Add an equal volume of phenol-chloroform-isoamyl alcohol (25:24:1). Mix properly for 5 min and centrifuge at 12,000×g for 10 min at RT. This step is performed to denature and remove RNAse.
18. Transfer the aqueous phase into a fresh tube using a wide-bore pipette (Fig. 1).
19. Reprecipitate DNA by adding 1/10 volume of 3 M sodium acetate and 2 volumes of chilled ethanol. Mix gently by inversion and let the DNA precipitate for 30 min at −20 °C (Fig. 2).
20. Spool out the precipitated DNA (Fig. 3) or centrifuge at 12,000×g for 15 min at RT and decant the supernatant.
21. Wash DNA pellet in 70% ethanol twice and air-dry till the ethanol evaporates.
22. Resuspend pellet in 100 μL TE buffer and store at −20 °C.

*Advantage*: Provides high-quality DNA which is appropriate for sequencing. Characteristics of high-quality DNA include (a) high molecular weight with little band shearing as accessed in 0.8% agarose gel, (b) A260/A280 between 1.8 and 2 which signifies purity of DNA from contamination by proteins and polyphenols, and (c) A260/A 230 between 1.5 and 1.8 which signifies purity of DNA from contamination by polysaccharides.

*Drawbacks*: However, this procedure suffers from the following three limitations: (a) The protocol is lengthy and time-consuming; (b) several washing and precipitation steps are involved, which results in the loss of DNA; and (c) the PCI treatment is not generally preferred due to the involvement of hazardous chemicals.

### 3.3.1 The Other Two Protocols with Some Variations Are Also Discussed Below

A. By Henry (1997)

Steps 1–8 remain almost the same with slight variation as per the need.

9. Add 1/10 volume of 7.5 M ammonium acetate and 2 volumes of chilled ethanol (−20 °C).

10. Mix by gentle inversion and keep at −20 °C for 1 h (DNA precipitates; Fig. 2).

11. Centrifuge at 10,000×g for 10 min and discard the supernatant. (RNA and chloroform are soluble in the supernatant.)

12. Wash the pellet twice with chilled 70% ethanol (washes the salts).

13. Air-dry the sample.

14. Resuspend and store in the TE buffer.

*Advantages*

- No RNase A treatment is required.
- Fewer steps.
- Avoids PCI treatment (as no need to remove RNase).

B. Extraction of sequencing quality DNA from recalcitrant plants (Healey et al., 2014):

- Steps 1–5 remain almost the same with slight variation as per the need.

6. Centrifuge at 5000×g for 5 min and transfer the aqueous phase into a fresh tube (Fig. 1).

7. Add 5 µL of RNase (10 µg/mL) and incubate in a 37 °C bath for 15 min with gentle mixing.

8. Add an equal volume of chloroform-isoamyl alcohol (24:1), and mix by gentle inversion for 5 min. (RNase gets removed in the organic phase; Fig. 1.)

9. Centrifuge at 5000×g for 5 min and transfer the aqueous phase into a fresh tube.

10. DNA precipitation: Add half volume of 5 M NaCl and mix gently by inversion; to this, add three volumes of chilled 95% ethanol. Place the tube in −20 °C for 1 h. (Don't exceed this time as CTAB and SDS can precipitate.)

11. Centrifuge at 5000×$g$ for 5 min and decant the supernatant.
12. Wash with 3 mL 70% ethanol and centrifuge at 5000 g for 10 min.
13. Air-dry the pellet, resuspend in the TE buffer, and store it at −20 °C.

*Advantages*

- Provide high quality of DNA.
- Sufficiently high quantities of DNA are obtained as the number of precipitation and washing steps are less.
- PCI treatment is not required.

# 4   Precautions

- Autoclave all the buffers/reagents for the protocol.
- Neither the plant tissue should be allowed to thaw before grinding, nor should the ground tissue in the mortar be allowed to thaw without the extraction buffer.
- CTAB must be kept in a water bath at 65 °C for some time before use. This is because CTAB with high salt concentration is very viscous and needs heating to make it flowy.
- PVP and ME must be added to the extraction buffer just before use.
- During phenol-chloroform-isoamyl alcohol extraction, TE buffer is used to maintain a neutral to slightly basic pH, as stability of DNA reduces at acidic pH.
- After phenol-chloroform-isoamyl alcohol extraction, remove the aqueous phase containing DNA using a wide-bore pipette in order to cause minimum damage to DNA strands.
- Ensure the removal of the aqueous phase without the interphase or organic phase components.
- Care should be taken to carry the protocol gently to avoid DNA shearing. Avoid vortexing.
- It is better to spool out the precipitated DNA as it avoids contamination.
- If spooling out of DNA is not possible, samples can be centrifuged. Be careful while discarding the supernatant as the pellet is very loose.
- Ensure that DNA never gets overdried during air-drying; otherwise, its resuspension in TE buffer will become problematic.
- Blank extraction controls are prepared with regular DNA extraction to check for any contamination.

# 5   Applications

Isolated DNA can be used for:

- *Polymerase chain reaction* (PCR amplification)

The extracted DNA cannot be directly analyzed. As it is the whole genomic DNA, the sequence of interest needs to be amplified before it can be used for sequencing or genetic fingerprinting. PCR serves as an acellular or molecular cloning method that amplifies the DNA of interest, producing millions of copies in the purest form. Once amplified, the detection and analysis of this DNA of interest can be done by agarose gel electrophoresis.

– *Sequencing*

The nucleotide sequence of the gene of interest, amplified by PCR, is determined by the dye-terminator sequencing technique, which is the advanced variant of Sanger's method. The sequence information is used for DNA barcoding of plants, which is basically a technique used for characterizing species using short DNA sequences from standard positions in the genome. This technique is employed in identifying a new plant with respect to a known classification (Arif et al., 2010).

– *Marker-assisted polymorphism detection*

The various molecular marker techniques like random amplified polymorphic DNA (RAPD), amplified fragment length polymorphism (AFLP), microsatellites or single sequence repeats (SSRs), and single nucleotide polymorphism (SNP) provide information on the distribution of variation in the species gene pool and help in understanding their evolutionary and taxonomic positioning. They also provide easy and reliable identification of plant species which is of great importance in the characterization and conservation of genetic diversity.

– *Next-gen sequencing* (NGS)

This non-Sanger-based sequencing method is swift and is capable of sequencing the complete genome rather than just a part of a gene. Hence, the entire isolated DNA can be taken for sequencing, and no PCR is required. It enables the sequencing of large numbers of samples at a lower cost. As rapid genome sequencing from various plant species can be done, it gives a better understanding of how genotypic variations translate into phenotypic characters. This gives a better perspective of plant evolution. Secondly, NGS plays an important role in crop improvement. Conventional plant breeding is dependent on the genetic variations available in the gene pool, which is determined by the phenotype. This approach has few drawbacks as it is highly time-consuming, and the genetic variation of interest may not be available in the narrow gene pool. By induced mutagenesis, plant breeders can rapidly produce mutant alleles that serve as a source of genetic variation for crop improvement with little effect on viability. NGS has enabled quick and easy identification, mapping, and characterization of these mutant genes with economically important traits (Sahu et al., 2020).

# References

Arif, I. A., Bakir, M. A., Khan, H. A., Al Farhan, A. H., Al Homaidan, A. A., Bahkali, A. H., Sadoon, M. A., & Shobrak, M. (2010). A brief review of molecular techniques to assess plant diversity. *International Journal of Molecular Sciences, 11*(5), 2079–2096. https://doi.org/10.3390/ijms11052079

Chiong, K. T., Damaj, M. B., Padilla, C. S., Avila, C. A., Pant, S. R., Mandadi, K. K., Ramos, N. R., Carvalho, D. E., & Mirkov, T. E. (2017). Reproducible genomic DNA preparation from diverse crop species for molecular genetic applications. *Plant Methods, 13*. https://doi.org/10.1186/s13007-017-0255-6

Doyle, J. J., & Doyle, J. L. (1987). A rapid DNA isolation procedure for small quantities of fresh leaf tissue. *Phytochemical Bulletin, 19*, 11–15.

Healey, A., Furtado, A., Cooper, T., & Henry, R. J. (2014). Protocol: A simple method for extracting next-generation sequencing quality genomic DNA from recalcitrant plant species. *Plant Methods, 10*. https://doi.org/10.1186/1746-4811-10-21

Henry, R. J. (1997). *Practical applications of plant molecular biology*. Chapman and Hall.

Jadhav, K. P., Raja, R. V., & Natesan, S. (2015). Chemistry of plant genomic DNA extraction protocol. *Bioinfolet, 12*, 543–548.

Sahu, P. K., Sao, R., Mondal, S., Vishwakarma, G., Gupta, S. K., Kumar, V., Singh, S., Sharma, D., & Das, B. K. (2020). Next generation sequencing based forward genetic approaches for identification and mapping of causal mutations in crop plants: A comprehensive review. *Plants (Basel, Switzerland), 9*(10). https://doi.org/10.3390/plants9101355

Tamari, F., Hinkley, C. S., & Ramprashad, N. (2013). A comparison of DNA extraction methods using *Petunia hybrida* tissues. *Journal of Biomolecular Techniques, 24*, 113–118. https://doi.org/10.7171/jbt.13-2403-001

Xia, Y., Chen, F., Du, Y., Liu, C., Bu, G., Xin, Y., & Liu, B. (2019). A modified SDS-based DNA extraction method from raw soybean. *Bioscience Reports, 39*(2). https://doi.org/10.1042/BSR20182271

# DNA Extraction by Spooling Method

**Abstract** A typical mammalian cell contains about six picograms and 2.2 m of DNA molecule. Therefore, the isolation of whole DNA from any human organ consisting of around 200 billion cells will be challenging and tedious. But the extraction of DNA by spooling method helps the experimenter isolate such a large quantity and long fibers of DNA conveniently. The nonexperts, in particular, can visualize the white translucent DNA fibers after spooling in this technique. The basic protocol consists of cell lysis and selective precipitation of DNA using different biochemicals. This is followed by the rapid spooling of precipitated DNA fibers on the glass rod or stirring rod under cold temperature conditions. Though the obtained DNA is free from proteins and other impurities, further processing of this DNA might be required for specific molecular biological applications.

**Keywords** DNA · Spooling · Fibers · Mammalian cells · RNAse

## 1 Introduction of the Technique

Though DNA from a specific cell can be isolated and analyzed with advanced protocols, it's not ample to visualize with the naked eye. To get enough DNA, millions of cells need to be lysed, and their DNA is spooled out of the solution. This protocol is known as DNA isolation by spooling method and is suitable for visualizing DNA strands from the mammalian cells (Sambrook & Russell, 2006; Zhang & Alvarado, 2018). The isolated DNA can be analyzed qualitatively and quantitatively, similar to other isolation methods. Due to the protocol's simplicity and fewer requirements of chemicals, spooling method of DNA isolation is widely used as a lab demonstration method in teaching schools and colleges.

## 2   Basic Principle

The DNA enclosed in the cell membrane and the nuclear membrane is taken out in the solution by a suitable cell lysis buffer. The addition of cell lysis buffer also provides a neutral charge to the DNA molecule because the positive ions of the buffer's salts bind to the negatively charged DNA backbone. The cold buffer and ethanol inhibit the action of nucleases and other degrading enzymes. Finally, DNA is precipitated out of the solution by adding cold polar solvent, i.e., ethanol (Fig. 1). This precipitated DNA is wound around a glass rod and taken out of the solution for further studies. This is called DNA spooling (Sambrook & Russell, 2006). DNA spooling can isolate DNA's long strands from a solution. More water is pushed out of DNA by spooling the strands tighter, and the DNA becomes silvery instead of fluffy white strands.

## 3   Protocol

### 3.1   Reagents Required

- 1 M Tris-HCl (pH 8.0)
- 0.5 M EDTA
- 10 M NaCl

**Fig. 1**  White DNA fibers are visible at the interphase of ethanol (*top*) and cell lysis buffer (*bottom*) in a test tube

- 10% SDS solution
- Deionized water
- Ethanol
- TE buffer (pH 8.0)
- RNase

## 3.2 Working Solutions

| Solution | Preparation | Storage |
|---|---|---|
| Cell Lysis Buffer (10 mM Tris-HCl, 0.5 M NaCl), 500 mL | Mix 5 mL of 1 M Tris-HCl (pH 8) and 1 mL 0.5 M EDTA to 450 mL deionized water. Add 5 mL of 10% SDS solution to it and 25 mL of 10 M NaCl. Make up the volume to 500 mL using deionized water | Room temperature |
| TE buffer (pH 8) 100 mL | Mix 1 mL Tris-HCl (pH 8) and 0.2. mL of 0.5 M EDTA in a glass beaker. Make up the volume to 100 mL by adding deionized distilled water | Room temperature |
| RNase A Solution (4 mg/mL) | Dissolve 10 mg RNAase in 2.5 mL TE buffer | 2–8 °C |
| Ethanol (ethyl alcohol), absolute alcohol | Use as purchased, better when used chilled | 2–8 °C |

## 3.3 Procedure

1. For cells, lyse them after making their suspension in the 7–8 volumes of cell lysis buffer in a polypropylene tube (Nath, 1990). For tissues, mince the tissue in a glass Petri dish, and spread the lysis buffer over the finely chopped tissue in 7–8 volumes of cell lysis buffer.
2. Transfer the entire content in a polypropylene tube or glass beaker and close it tightly. Keep it at room temperature for 1 h.
3. With the help of wide-bore tips, put layers of ~20 mL chilled ethanol through the sides of the tube or beaker. Avoid mixing with the content.
4. Swirl the mixture in one direction using a glass rod to form interphase between cell lysate and ethanol.
5. Remove the white strands of DNA carefully from the interphase by glass rod or Shepherd's crook (Fig. 1).
6. Keep and store the isolated DNA strands in 1–5 mL of ethanol in a fresh tube.
7. Before using this isolated DNA, rehydrate the DNA in another tube with 1 mL TE buffer (pH 8) overnight at 4 °C.
8. Finally, purify the DNA by removing RNA contamination by adding 1 μg/mL RNase to the extracted DNA.

# 4   Precautions

- Recovery of DNA should be made by slow stirring only.
- Dispose of used reagents according to local ordinances.
- Wear latex gloves during the procedure.
- The laboratory surfaces should be spotless during all procedures used in this activity.
- Use caution while using highly flammable ethanol.

# 5   Applications

Depending upon the purification and digestion steps followed after the isolation of DNA, the isolated DNA can be used for any application. This includes applications that require the DNA in a large amount (e.g., formation of DNA libraries and NGS) or DNA from a gross tissue or organ (e.g., in medical diagnosis and genomics) (Zhang & Alvarado, 2018).

# References

Nath, K. (1990). A rapid DNA isolation procedure from petri dish grown clinical bacterial isolates. *Nucleic Acids Research, 18*(21), 6462.

Sambrook, J., & Russell, D. W. (2006). Isolation of DNA from mammalian cells by spooling. *Cold Spring Harbor Protocols, 2006*(1), pdb-prot4037.

Zhang, S., & Alvarado, A. S. (2018). Planarian high molecular weight DNA isolation by spooling. In J. Rink (Ed.), *Planarian regeneration. Methods in molecular biology* (Vol. 1774). Humana Press.

# Magnetic Bead-Based Nucleic Acid Isolation

**Abstract** Magnetic bead-based techniques are majorly used to extract and purify the nucleic acids directly from all types of crude samples like tissues of blood, hair or nails, soil, fungal growth, and viruses. It is advantageous over other extraction strategies such as cost-effective, rapid, eco-friendly, and high yield. This strategy is based upon the principle of paramagnetism or superparamagnetism. An external magnetic field is applied to a solution containing these tiny nanoparticles/beads, which behaves as small paramagnets due to induced spin. The efficiency and diversity of magnetic beads to extract biological samples are enhanced by enriching their surface chemistry by functionalizing with appropriate chemicals. The major processing steps include binding target biomolecule with beads, immobilization through applying external magnetic field, washing, and extraction of nucleic acid from beads. These techniques possess extensive applications in genetic engineering, hereditary and genomic identification, diagnosis, pharmaceuticals, and vaccine development. However, the next-generation magnetic bead separation techniques are intelligent and automated, requiring specific processing with the Internet of things.

**Keywords** Magnetic separation · Magnetic beads · Paramagnetism · Molecular spin · Nucleic acid separation · Crude sampling · Surface functionalization

## 1 Introduction

The potential of magnetic beads for nucleic acid separation and purification was recognized in the 1990s with a publication of US patent on "DNA purification and isolation using magnetic particles" (Hawkins, 1998). The use of magnetic beads for nucleic acid isolation possesses various advantages over other techniques, including high yield, economical, user-friendly nature, the possibility of automated process, and lesser chance of cross-contamination due to the exclusion of harmful chemicals and significant scalability (Chen, 2020). Due to the use of functionalized beads, this technique provides the adequate lysis of the tissue and cells, nucleoprotein

© The Author(s), under exclusive license to Springer Nature Switzerland AG 2022    111
A. Gautam, *DNA and RNA Isolation Techniques for Non-Experts*, Techniques in Life Science and Biomedicine for the Non-Expert,
https://doi.org/10.1007/978-3-030-94230-4_15

denaturation, and a nuclease-free environment (Kovačević, 2016). The fascinating feature of this technique is that the nucleic acid can be isolated directly from the crude samples, including blood, tissues from hairs or nails, homogenates, cultivation media, soil, amalgam loom, fungal species, and viruses. The machinery required is exceedingly cost-effective in manufacturing and being installed in a laboratory setting (Kovačević, 2016; Ali, 2017).

Magnetic beads are small nano- or micro-sized particles studied as potential ways to solve scalability issues regarding nucleic acid isolation, extraction, and manipulation (Berensmeier, 2006). These beads can be quickly immobilized and resuspended by applying alternate magnetic fields, making them a promising candidate for nucleic acid extraction. They can be integrated with an appropriate buffer system, which allows fast and efficient purification and the extraction of DNA/RNA from complex cell extracts or other biological tissues. Protocols for magnetic beads are implicitly scalable, independent of tedious centrifugation (Obino, 2021). The use of such small particles ensures high chemical stability with lesser particle-particle interaction (Kim, 2020).

Since the interaction of these beads with DNA is a surface activity, the smaller size of particles leads to a larger surface-to-volume ratio, which enhances the surface interactions. Metals such as iron, cobalt, and nickel-based magnetic particles are the most used in biomedical and genomic applications (Tang, 2020). Various metal oxides magnetite $Fe_3O_4$ and maghemite $\gamma$-$Fe_2O_3$ are the most popular among these magnetic particles for nucleic acid and immunological assays because of their excellent biocompatibility, faster separation, and significant stability under external magnetic fields (Vanyorek, 2021). These enhanced characteristics make bead-based RNA/DNA separation method automation and user-friendly and a promising suitable candidate for diversified applications, including sample preparation for next-generation sequencing (NGS), immunodiagnostics, protein purification, and PCR analysis.

## 2  Basic Principle

Magnetic separation is a technique used to separate paramagnetic or superparamagnetic particles (of size less than the macroscale range) from a suspension by applying an external magnetic field. Due to induced spin, paramagnetic particles act like tiny magnets and interact with an externally applied magnetic field (Akbarzadeh, 2012). The interaction between these induced magnetic particles and an applied magnetic field is the basic principle of magnetic separation. In genetic engineering and molecular biology, magnetic separation purifies different biomolecules, including plasmids, mitochondrial DNA, genomic DNA, RNA, and proteins (Kim, 2020). A particular type of bead with the required appropriate chemistries and coating is used to separate specific target biomolecules, as listed in Table 1. Such beads lead to direct extraction of desired target biomolecule from the crude sample with

**Table 1**  Different types of beads and their applications

| Type | Applications |
|---|---|
| Carboxylate-modified magnetic beads | • Affinity to directly capture DNA/RNA<br>• Conjugate through a covalent bond |
| Streptavidin-coated magnetic beads | • Able to bind bio-ligands such as DNA/RNA<br>• Fast reaction kinetics<br>• High precision |
| NeutrAvidin-coated magnetic beads | • Alternative to streptavidin<br>• Properties similar to streptavidin |
| Protein A/G magnetic beads | • Able to bind IgG and IgA proteins<br>• Vast binding properties<br>• Immunoprecipitation |
| Silica-coated magnetic beads | • Based on salt concentration reversibly binds DNA/RNA<br>• Monodispersion of small-sized particles in a range of few hundred micrometers<br>• Application in PCR |

minimal processing. Thus, the principle of this technique involves two basic principles: paramagnetism and surface chemistries of functionalized beads (Oberacker, 2019).

The process of extraction can be broadly divided into four stages as illustrated in Fig. 1:

Stage 1: Binding of target DNA to magnetic bead—In the first stage, magnetic beads with specific coatings and chemistry are added to the sample containing target molecules, which binds the target DNA from the sample by adjusting its buffer conditions.

Stage 2: Immobilization through magnetic separation—In the second stage, an external magnetic field is applied to the sample, attracting the beads toward the outer edge of the vessel containing the sample and immobilizing them.

Stage 3: Washing of sample with immobilized beads—In the third stage, the sample is washed. The beads with immobilized DNA retains during washing due to the application of an external magnetic field.

Stage 4: Release of DNA from beads for further analysis—In the last stage, after removing the externally applied magnetic field, a stabilizing buffer solution such as elution buffer to beads releases the captured DNA as a purified sample, which can be further used for various analyses and applications. Thus, the use of magnetic beads eliminates the requirement of vacuum or centrifugation for the extraction process, thereby reducing the extraction cost and increasing its popularity among the scientific community.

These days all the stages are automated through commercially available apparatus. Thus, these are automated methods for nucleic acid separation based on magnetic separation.

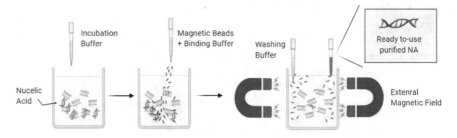

**Fig. 1** Basic steps of magnetic bead-based isolation of nucleic acid

# 3 Protocol

## 3.1 Reagents Required

- 1% (w/v) sodium dodecyl sulfate (SDS)
- Magnetic beads—as supplied by the manufacturer
- 10 M NaCl
- 10% polyethylene glycol (PEG) 6000—prepared in 0.1 M Tris-HCl (pH 8.0)
- Ethanol
- Deionized water
- 1 M Tris base (pH 10–11)
- 0.5 M EDTA

## 3.2 Working Solution

| Solution | Preparation | Storage |
|---|---|---|
| Incubation buffer | 30 µL of 1% (w/v) SDS solution for 30 µL of sample | RT |
| Binding buffer (100 mL) | Add 10 mL 0f PEG 6000 in 12.5 mL of 10 M NaCl. Make up the volume to 100 mL by adding deionized water | RT |
| Elution buffer or TE buffer (100 mL) | Add 1 mL of 1 M Tris base (pH 10–11) and 0.2 mL EDTA (0.5 M). Make up the volume to 100 mL with deionized water | RT |
| Wash buffer (100 mL) | Dilute 70 mL absolute alcohol (ethanol) with 30 mL of deionized water | RT |

## 3.3 Procedure

The following method has been adapted from the protocol used by Saiyed (2003) to extract nucleic acids using bare magnetic beads:

1. For a typical extraction process, add 30 μL of 1% (w/v) SDS solution and the same amount of sample containing the target DNA molecule in a microfuge tube/vessel.
2. After addition, invert the microfuge tube/vessel several times to mix the solution, and then incubate for a minute at room temperature for lysis of the cell.
3. Upon incubation, add 10 μL of bare magnetic nanoparticles/beads and 75 μL of binding buffer to this solution.
4. Mix the obtained suspension thoroughly by inversion, and keep it at room temperature for 3 min. This will help the binding of nucleic acid to the beads.
5. Using an electromagnet, apply an external magnetic field to immobilize the magnetic pellet (DNA/RNA with beads).
6. After removing the externally applied magnetic field, wash the obtained magnetic pellet using 70% ethanol and air-dry it.
7. Resuspend this pellet in 50 μL of TE buffer and elute the DNA/RNA from magnetic beads by incubation in automated apparatus or water bath at +65 °C with continuous agitation.
8. If there are magnetic beads carried over with extracted DNA/RNA, again apply the magnetic field over eluant tube to remove the residual beads and transfer the eluant into another new tube.
9. The purified RNA/DNA can be used for downstream applications or can be stored at −20 ° C.

This procedure yields a sufficient amount of DNA, which can be used for at least 30 PCR reactions. However, these days automated apparatus based on magnetic separation are available in the market, which provides a better yield of DNA/RNA.

## 4   Precautions

The following precautionary measures must be taken to get a high yield of DNA/RNA:

- The surface of magnetic beads must be free from any contaminants, nuclease, protease, and enzymes.
- It is better to minimize the operating steps to avoid anomalies.
- The magnetic beads taken must be of uniform size and must not dissolve in solution.
- Surface functionalized magnetic beads chosen must not indulge chemically or initiate any chemical reaction during the extraction process.
- The magnetic field must be applied uniformly and must be varied gradually.

# 5   Applications

Genomic and hereditary studies: Amplification and cloning of the isolated nucleic acid using PCR and cloning vector techniques provide a larger sample size for experimentation. Further computational and in silico studies can be carried out for understanding and identifying patterns in the sequence, the binding capacity of ligands, and molecular interaction (Lee, 2020).

Diagnostics: Many medical conditions can often be diagnosed at a controllable stage with the help of DNA/RNA extracted from a patient. Such conditions include cystic fibrosis, fragile X syndrome, hemophilia A, sickle-cell anemia, Down's syndrome, Huntington's disease, and Tay-Sachs disease, which can be diagnosed by genetic testing. Besides diagnosing existing diseases, geneticists also speculate whether a person is a potential carrier of a recessive trait without the disease (Oberacker, 2019). Recently, magnetic bead separation-based nucleic acid extraction has also shown potential in virology, especially coronavirus disease (COVID-19) (Smyrlaki, 2020).

Pharmaceuticals: Nucleic acid extraction is regarded as the initial step in the manufacturing and trials of several pharmaceuticals. Pharmaceuticals made via recombinant genetics include the synthesized insulin and human growth hormone (HGH) (Richardson, 2018).

Vaccines: The nucleic acid vaccine appears to be one of the most promising steps in the field of applied gene therapy. With the ability to readily induce humoral and cellular immune responses, it has gained massive traction. A DNA/RNA vaccine is considered to be potentially safer than the traditional vaccination being followed. Apart from being more stable, it is potentially more cost-effective to manufacture and for storage purposes. Diseases targeted are malaria, hepatitis B, COVID-19, etc. (Wommer, 2021).

Forensic science: DNA extraction is used in forensic sciences for various purposes such as fingerprinting, identifying the criminals by matching the DNA of the accused obtained from the crime site, and identifying parenting (Dierig, 2020).

# References

Akbarzadeh, A. S. (2012). Magnetic nanoparticles: Preparation, physical properties, and applications in biomedicine. *Nanoscale Research Letters, 7*, 144.

Ali, N. R. (2017). Current nucleic acid extraction methods and their implications to point-of-care diagnostics. *BioMed Research International, 9306564*.

Berensmeier, S. (2006). Magnetic particles for the separation and purification of nucleic acids. *Applied Microbiology and Biotechnology, 73*(3), 495–504.

Chen, Y. L. (2020). Magnetic particles for integrated nucleic acid purification, amplification and detection without pipetting. *Trends in analytical chemistry: TRAC, 127*, 115912.

Dierig, L. S. (2020). Looking for the pinpoint: Optimizing identification, recovery and DNA extraction of micro traces in forensic casework. *Forensic Science International: Genetics, 44*, 102191.

Hawkins, T. (1998). *Patent No. 5,705,628.*

Kim, S.-E. T.-H. (2020). Magnetic particles: Their applications from sample preparations to bio-sensing platforms. *Micromachines, 11*, 302.

Kovačević, N. (2016). Magnetic beads based nucleic acid purification for molecular biology applications. In M. Micic (Ed.), *Sample preparation techniques for soil, plant, and animal samples.* Springer Protocols Handbooks.

Lee, K. T. (2020). Parallel DNA extraction from whole blood for rapid sample generation in genetic epidemiological studies. *Frontiers in Genetics, 11*, 374.

Oberacker, P. S. (2019). Bio-on-magnetic-beads (BOMB): Open platform for high-throughput nucleic acid extraction and manipulation. *PLoS Biology, 17*(1), e3000107.

Obino, D. V. (2021). An overview on microfluidic Systems for Nucleic Acids Extraction from human raw samples. *Sensors (Basel), 21*, 3058.

Richardson, D. I. (2018). Accelerated pharmaceutical protein development with integrated cell free expression, purification, and bioconjugation. *Scientific Reports, 8*, 11967.

Saiyed, Z. T. (2003). Application of magnetic techniques in the field of drug discovery and bio-medicine. *BioMag Res Tech, 1*, 2.

Smyrlaki, I. E. (2020). Massive and rapid COVID-19 testing is feasible by extraction-free SARS-CoV-2 RT-PCR. *Nature Communications, 11*, 4812.

Tang, C. H. (2020). Application of magnetic nanoparticles in nucleic acid detection. *Journal of Nanbiotechnology, 18*, 62.

Vanyorek, L. I.-D. (2021). Synthesis of iron oxide nanoparticles for DNA purification. *Journal of Dispersion Science and Technology, 42*(5), 693–700.

Wommer, L. M. (2021). Development of a 3D-printed single-use separation chamber for use in mRNA-based vaccine production with magnetic microparticles. *Engineering in Life Sciences*, 1–16.

# Density Gradient-Based Nucleic Acid Isolation

**Abstract** Nucleic acid (DNA/RNA) isolation has a wide range of applications, right from research purposes to applications such as food development, identification of the new organism and their categorization, diagnosing patients with genetic mutations, and also the expression levels of specific genes. Therefore, the scientific fraternity has devised new and innovative methods for the isolation of the said biomolecules. However, although a little old, the density gradient method is yet being used in many laboratories and yields decent results. The method employs a very simple principle in which the components are separated on the basis of their density with respect to the solvent used. Cesium chloride (CsCl) is one of the popular solvents used in the density gradient method. Under the influence of high centrifugal force, a solution of cesium chloride (CsCl) molecules will dissociate, leading to the formation of a narrow density gradient. The nucleic acid in question will migrate in this gradient to the point with a similar density (the neutral buoyancy or isopycnic point). This present article gives a detailed view of the isolation of the nucleic acids using the density gradient method. It highlights the different steps and the precautions one needs to observe while working on the said experiment.

**Keywords** Nucleic acids · Deoxyribonucleic acid · Ribonucleic acid · Density · Cesium chloride

## 1 Introduction of the Technique

Density gradient centrifugation or ultracentrifugation is one of the common techniques used to isolate and purify biomolecules and cell structures. It can simultaneously isolate intact DNA-free RNA, genomic DNA, and proteins from a biological specimen which can be useful in cloning genes and analyzing gene expression (Zhang et al., 2003). CsCl (cesium chloride), when used in the density gradient method, becomes a valuable means for fractionating, quantitatively separating, and

© The Author(s), under exclusive license to Springer Nature Switzerland AG 2022
A. Gautam, *DNA and RNA Isolation Techniques for Non-Experts*, Techniques in
Life Science and Biomedicine for the Non-Expert,
https://doi.org/10.1007/978-3-030-94230-4_16

**Fig. 1** Visualization of
separated bands of DNA
under blue light
illumination after
ultracentrifugation

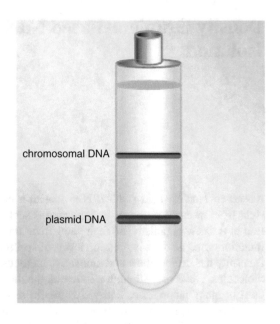

characterizing RNA and DNA, and the total quota of proteins, respectively, based
on differences in their buoyant densities (Taulbee & Furst, 2005) (Fig. 1).

## 2   Basic Principle

Density gradient centrifugation exploits the fact that, in suspension, particles having
a higher density than the solvent will sediment, while those that have less density
than the solvent will float. A high-speed ultracentrifuge can be employed to quicken
this process to separate biomolecules within a density gradient, which can be established by layering liquids of decreasing density in a centrifuge tube (JoVE Science
Education Database, 2021).

The spinning renders sample solutions present in containers (tube or small bot-
tles) to experience a centrifugal force resulting in the samples being pushed away
from the center of the rotor toward the bottom of the tube. Centrifugal force causes
the separation of the sample components based on their size because the larger components experience greater centrifugal force than smaller components. Thus, centrifugal force drives the larger components of a mixture farther from the rotor and
closer to the bottom of the tube (Taulbee & Furst, 2005).

In principle, density gradient centrifugation can be subcategorized into two key
types: rate zonal and isopycnic. Briefly, in isopycnic density gradient centrifugation,
a high-density gradient is used, and cells are separated exclusively on differences in
density. In contrast, a lower-density gradient is employed in rate zonal density gradient centrifugation, and cells are predominantly separated on size differences
(Burdon et al., 1988).

# 3 Protocol

## 3.1 Reagents Required

- Lysozyme (125 mg in 1000 μL TE buffer)
- RNase A (10 μg/mL)
- Proteinase K
- 20% SDS
- Phenol
- Chloroform
- Isoamyl alcohol
- Tris base
- HCl
- EDTA
- Distilled water
- CsCl (cesium chloride)
- Ethidium bromide (EtBr)

## 3.2 Working Solutions

| Solutions | Preparation | Storage |
|---|---|---|
| Lysozyme (1 mL) | Dissolve 25 mg lysozyme powder in 1 mL of 10 mM Tris-Cl, pH 8.0 at 37 °C. Thaw before use | −20 °C |
| RNase A (2.5 mL) | Dissolve 10 mg RNAase in 2.5 mL TE buffer | −20 °C |
| Proteinase K (1 mL) | 20 mg/mL in Tris-HCl (pH 8). Thaw before use | −20 °C |
| 20% SDS (100 mL) | Add 20 g SDS in 100 mL distilled water and stir slowly | RT |
| Phenol-chloroform-isoamyl alcohol (25:24:1, saturated with 10 mM Tris, pH 8.0, 1 mM EDTA) | Use as purchased, no preparation needed | 2–8 °C |
| Chloroform-isoamyl alcohol (49:1) (100 mL) | Mix 98 mL chloroform and 2 mL isoamyl alcohol (3-methyl-1-butanol) in a glass beaker | RT |
| Isopropanol | Use as purchased; no preparation needed | RT |
| TE buffer (pH 8, 100 mL) | Mix 10 mL 1 M Tris buffer and 1 mL 0.5 M EDTA in 75 mL distilled water. Maintain pH 8 and make up volume to 100 mL by adding more water | RT |
| Tris-HCl | Take 1 mL of 1 M Tris-HCl (pH 8.0) and add 99 mL of deionized water | RT |
| 0.5 M EDTA (pH 8, 100 mL) | Mix 14.61 g EDTA in 95 mL distilled water. Maintain pH by NaOH and makeup volume to 100 mL by adding more water | RT |

## 3.3 Protocol

1. Tissue or cells should be incubated with 10 µL lysozyme and 5 µL RNase A for 1 h at 37 °C. Add 100 µL of Proteinase K and 100 µL of 20% SDS for the next incubation, and place the mixture in an incubator at 55 °C for 1–2 h. Transfer the lysate into a 15 mL Falcon tube.
2. Add an equivalent amount (about 3 mL) of phenol-chloroform-isoamyl alcohol to the lysate tube. Vortex for 10 seconds to provide a thorough mixing.
3. Keep the tube for centrifugation at 3000×g for 5 min. Centrifuging the phenol-chloroform-lysate mixture should result in the separation of the mixture into two layers: an organic layer on the bottom that contains proteins from the sample and an aqueous layer on the top that contains DNA and other more hydrophilic molecules.
4. Transfer the aqueous layer (which contains DNA) into a fresh 15 mL Falcon tube.
5. Add about 3 mL of chloroform-isoamyl alcohol to the new tube and vortex briefly. This procedure should remove any phenol that may have remained in the DNA sample.
6. Centrifuge at 3000×g for 5 min or until the aqueous layer is clear. Transfer the aqueous layer to a fresh Falcon tube and add an equivalent quantity of isopropanol.
7. Remove the supernatant after centrifuge at 10,000×g for 5 min.
8. Wash the DNA pellet with TE buffer thrice. Add 500 µL of TE buffer and store the DNA at 4 °C (Somerville et al., 1989).

   Note: *By running samples over a 0.8% agarose gel electrophoresis, determine the DNA quality and estimate the quantity of the DNA in samples. The purification step can be uptaken if the DNA quality and quantity are satisfactory using cesium chloride (CsCl) gradient centrifugation procedure.*

9. Transfer 500 µL of genomic DNA to a centrifuge tube with 0.5 g of CsCl. Put a tiny piece of parafilm on top of the tube, and gently invert the tube 10–20 times to ensure even distribution. Mixing with a pipette may cause DNA to be sheared; therefore, it should be avoided. Afterward, add 10 µL of 10 µg/µL EtBr to this tube, and stir it in the same manner as before.
10. Run at 200,000×g for 18 h at 20 °C.
11. Remove the tubes from the rotor and observe in a dark environment. DNA will be seen using a blue light transilluminator, or if one is not available, use long-wavelength UV light to visualize the DNA.
12. The last step is removing the DNA band and placing it in a 1.5 mL Eppendorf tube using a sterile 1 cc syringe and needle (26G 5/8).

# 4  Precautions

- CsCl has been reported to have mild toxicity; therefore, one should be careful while using it.
- Use a different pipette tip for each addition of phenol-chloroform-isoamyl alcohol to prevent cross-contamination.
- Utmost care should be taken while using EtBr since it is considered a mutagen and a probable carcinogen.
- The proper balance of tubes should be maintained while operating the centrifuge.
- Careful pipetting should be observed when transferring the aqueous layer from the phenol-chloroform solution to avoid contamination.

# 5  Applications

- This method can be primarily used for the isolation of intact DNA-free RNA, genomic DNA, and proteins from different biological samples.
- The genetic material obtained can further be used for the cloning of genes and gene sequencing.

# References

Burdon, R. H., van Knippenberg, P. H., & Sharpe, P. T. (1988). Centrifugation. *Laboratory Techniques in Biochemistry and Molecular Biology Elsevier, 18*, 18–69.

JoVE Science Education Database. (2021). *Biochemistry*. Density Gradient Ultracentrifugation. JoVE.

Somerville, C. C., Knight, I. T., Straube, W. L., & Colwell, R. R. (1989). Simple, rapid method for direct isolation of nucleic acids from aquatic environments. *Applied and Environmental Microbiology, 55*(3), 548–554.

Taulbee, D. N., & Furst, A. (2005). CENTRIFUGATION| Preparative.

Zhang, H., Chen, H. T., & Glisin, V. (2003). Isolation of DNA-free RNA, DNA, and proteins by cesium trifluoroacetate centrifugation. *Biochemical and Biophysical Research Communications, 312*(1), 131–137.

# DNA Extraction from Agarose Gel Through Paper Strips

**Abstract**  Be it forensic sciences, drug development, disease diagnosis, or mutation studies, all are related to DNA isolation. There are multiple methods through which DNA isolation can be done. Some techniques claim to provide highly pure DNA but trade-off with the yield, and others high at the yield but compromise purity. The technique of DNA extraction from agarose gel through paper strips is advantageous over its counterparts in the low cost and less instrumentation required. The technique works on a simple principle of inserting the filter papers between the slits cut on the agarose gel. Electrophoresis is continued until the bands of the DNA are not collected on the filter paper. This is followed by centrifugation for the elution of the DNA from the filter paper. The present article highlights the steps involved in this process and gives an overview of the precautions that should be kept in mind while performing this technique.

**Keywords**  Deoxyribonucleic acid · Agarose · Paper strips · DNA isolation

## 1   Introduction

The extraction of DNA from agarose gel using paper strips is a conventional and cost-effective technique. DNA obtained from apparently any source which can be run on a standard agarose gel can be isolated using this technique (Oliveira et al., 2014). DNA is first to run on the agarose gel and then extracted out of the agarose gel, which may bring together many impurities such as unspecified carbohydrates. Though these impurities can easily be removed using modern-day techniques, there is a trade-off between the recovery and the purity of the DNA sample (Sirakov, 2016; Vemuri et al., 2016). Methods characterizing extraction with organic solvents, electroelution, silica particles, or ion-exchange resins offer quite pure DNA at the cost of relatively low yields (Hengen, 1994). Contrarily, high-yield techniques pose

© The Author(s), under exclusive license to Springer Nature Switzerland AG 2022    125
A. Gautam, *DNA and RNA Isolation Techniques for Non-Experts*, Techniques in
Life Science and Biomedicine for the Non-Expert,
https://doi.org/10.1007/978-3-030-94230-4_17

problems in the enzymatic reaction due to the lack of purity. The most important feature that gives this technique an edge over the others is its low cost and less equipment-dependent approach. Therefore, it can be used in most scientific laboratories where educating undergraduate and postgraduate students is the sole aim (Yu & Morrison, 2004; Zou et al., 2017).

# 2    Basic Principle

This technique was firstly reported in the year 1980 by Girvitz and coworkers (Girvitz et al., 1980). The basic principle of the technique is that the DNA fragments can be recovered by inserting the strips of filter paper backed by dialysis membrane into the slits cut in the gel in front of the DNA and continuing electrophoresis until the DNA is collected in the paper. The DNA fragments that are smaller in size have the highest speed of migration. Furthermore, the higher is the percentage of the gel, the lower is the pore size. Henceforth, depending upon the size of DNA to be fractioned, the percentage of the gel is used. For instance, 0.8% gel is used to resolve a 5–10 kb fragments, 2% for 0.2–0.1 kb fragments, and 4% for 0.04–0.9 kb fragments. For resolving genomic DNA, 0.8% agarose gel; for resolving RNA and plasmids 1% agarose gel; for routine PCR amplified product, 2% agarose gel; and for ARMS PCR, 4% agarose gel is generally used. Low-speed centrifugation is further used for the elution of the DNA from the filter paper. A high recovery (above 70%) has been reported with this method. The recovered DNA is biologically active and can be ligated, labeled, and re-cleaved under in vitro conditions. In 1990, Errington reported a modification of Girvitz et al. (1980), which was a simplified version of the earlier method but worked on the same principle.

# 3    Protocol

## 3.1    Reagent Required

- 1 M Tris (pH 7.8)
- 3 M sodium acetate
- 0.5 M EDTA (pH 8.0)
- 10 mg/mL ethidium bromide
- 2% agarose gel
- 5X TBE

## 3.2   Working Solutions

| Working | Preparation | Storage |
|---|---|---|
| 40 mM Tris (pH 7.8) | Dissolve 30.285 g Tris in 100 mL DW (distilled water). Adjust pH to 7.8 and volume to 250 mL. Autoclave the solution | RT (room temperature) |
| 5 mM sodium acetate | Dissolve 61.5 g of sodium acetate in 100 mL of DW. Adjust volume to 250 mL. Filter sterilize and autoclave the solution | RT |
| 2 mM EDTA (pH 8.0) | Dissolve 46.5 g of EDTA in 100 mL of DW. Adjust the pH with NaOH pellets or 10 N NaOH. Make up volume to 250 mL. Filter, sterilize and autoclave the solution [approximately 20 g of NaOH pellet is required for 1 L of the solution to bring the pH to 8] | RT |
| 0.5 µg/mL ethidium bromide | Dissolve 2.5 g of ethidium bromide in 100 mL of autoclaved deionized water. Make up volume to 250 mL by adding more deionized water. Stir on a magnetic stirrer for several hours so that the dye dissolves completely. Store in a dark bottle or cover the bottle with aluminum foil | RT For long-term storage, the stock can be stored at 4 °C |
| 0.5X TBE (45 mM Tris-borate/1 mM EDTA) | Dissolve 13.5 g of Tris base, 6.875 g of boric acid in 100 mL of deionized water. Add 4 mL of 0.5 M EDTA. Adjust pH to 8.4 if required. Make up volume to 250 mL | RT |

### 3.2.1   Preparation of 2% Agarose Gel

Dissolve 2 g of agarose in 100 mL of 0.5X TBE. Heat until the solution boils, allow it to cool for a while, and add EtBr (0.5 µg/mL). Pour the solution into the gel casting tray, and immediately insert combs so that wells can be created in the gel to load the sample, and allow it to solidify. Once the gel has been solidified, pour some 0.5X TBE buffer near the wells and gently remove the combs. The gel is now ready to use.

### 3.2.2   Restriction Digestion Reaction

The following table shows a standard reaction plan for restriction digestion. The enzyme is always added in the end while setting the reaction to avoid any degradation of the enzyme. Once everything is mixed, the reaction tube is incubated for 1 h at 37 °C, and then to stop the reaction, either the incubation is followed by a further incubation at 65 ° C for 15 min or an addition of 10 mM EDTA to the reaction.

| Reagent | Volume (µL) |
|---|---|
| Autoclaved deionized water | 2.00 |
| Enzyme (1 U/µL)<br>How many units of enzyme should be ideally used for digestion varies from sample to sample. It can be standardized using different units of enzyme and keeping the substrate or sample concentration the same for all the units and further visualizing at which unit complete digestion has occurred<br>(mostly 2.5–5.0 units are used) | 5.00 |
| Enzyme buffer (10X) | 1.00 |
| Sample to be digested (PCR product, plasmid, genomic DNA) | 2.00 (total 1 µg final concentration, from 500 ng/µL concentration of the sample to be digested) |
| 10 mM EDTA | Used to stop the enzymatic reaction |

## 3.3    Procedure

As per the protocol given by Girvitz et al. (1980) and Errington (1990):

1. Samples are digested generally by restriction enzymes such as *EcoRI*.
2. Digested samples are run on the ordinary agarose gel, which can be used for restriction mapping (Fig. 1). Until sufficient separation of the DNA fragments occurs, the electrophoresis is continued at 1–2 V/cm.
3. Recovery of DNA from gel: A slit is carefully cut in the gel positioned right in front of the band corresponding to DNA to be removed, and in that slit, a strip of sterile Whatman 3 mm paper is inserted.
4. Following this, the gel is returned to the tank, and a short duration of electrophoresis is used to transfer the DNA into the paper (Fig. 2). Note that this step requires a lot of standardization and precaution, which have been mentioned under the precaution subheading.
5. If, due to some reason, the band passes through the paper, the procedure would again be repeated, starting from making a slit in the gel and placing the sterile 3 mm Whatman filter paper in the slit as previously done.
6. When the DNA band is partially trapped inside the paper, the paper is removed from the gel, and the excess of the paper which was protruding out from the gel is trimmed or cut.
7. The isolated paper from the gel is placed in a 500 µL microcentrifuge tube with a small hole in the bottom (hotwire can be used to pierce through the tube and form a hole).
8. This tube is placed inside a larger tube (1.5 mL tube) and microfuge for 20 s. This leads to the recovery of DNA in a small volume of electrophoresis buffer.

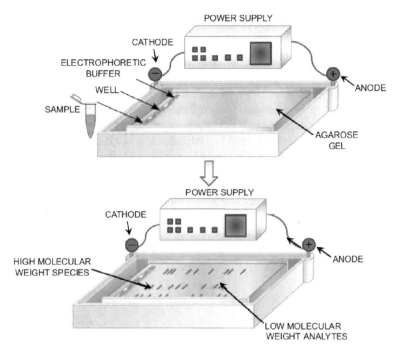

**Fig. 1** Agarose electrophoresis system. (Adapted from Drabik et al., 2016)

**Fig. 2** A paper strip inserted in the gel slit for the extraction of DNA onto the paper strip (BiochemPages, 2017)

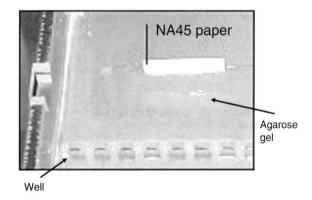

Usually, from a band of around 100 ng, 30–40 ng of DNA is obtained, which is sufficient for most cloning purposes. It is interesting to note here that loading more DNA would not lead to improved recovery. However, if several strips of paper are inserted in parallel, the recovery can be improved substantially.

9. Once the DNA is recovered, no other method is needed to purify it.

## 4    Precautions

- The strip inserted into the gel should be of the same width as that of the band but may protrude from the gel by a few mm.
- After inserting the paper into the gel, the electrophoresis duration should be properly standardized because electrophoresis for a longer duration would lead the DNA to pass right through the paper. Paper has mobility similar to that of a 0.7% agarose gel routinely used in molecular biology labs. Therefore, the time for which the electrophoresis runs can be calculated if the rate of the mobility of the fragment is known (calculated by the distance traveled by the band and the speed of the movement).
- For samples containing a range of sizes of the fragment, say from a partial DNA digest with a restriction enzyme, the paper can be inserted at an angle to the direction of fragment migration.
- Phenol extraction and precipitation steps decrease the yield, reliability, and speed of the method.

## 5    Applications

The DNA obtained from this method is biologically active and can be used for labeling, cleaving, and hybridization studies. Furthermore, the DNA obtained from this method can be directly used in the cloning and subcloning methods.

## References

BiochemPages. (2017). *Conventional methods for the agarose gel extraction of DNA*. Retrieved March 11, 2021, from https://www.biochempages.com/2017/04/conventional-methods-aga-rose-gel-extraction-dna.html

Drabik, A., Bodzoń-Kułakowska, A., & Silberring, J. (2016). Gel electrophoresis. In *Proteomic profiling and analytical chemistry* (pp. 115–143). Elsevier.

Errington, J. (1990). A rapid and reliable one-step method for isolating DNA fragments from agarose gels. *Nucleic Acids Research, 18*(17), 5324.

Girvitz, S. C., Bacchetti, S., Rainbow, A. J., & Graham, F. L. (1980). A rapid and efficient procedure for the purification of DNA from agarose gels. *Analytical Biochemistry, 106*(2), 492–496.

Hengen, P. N. (1994). Methods and reagents: Recovering DNA from agarose gels. *Trends in Biochemical Sciences, 19*(9), 388–389.

Oliveira, C. F. D., Paim, T. G. D. S., Reiter, K. C., Rieger, A., & D'azevedo, P. A. (2014). Evaluation of four different DNA extraction methods in coagulase-negative staphylococci clinical isolates. *Revista do Instituto de Medicina Tropical de São Paulo, 56*(1), 29–33.

Sirakov, I. N. (2016). Nucleic acid isolation and downstream applications. *Nucleic Acids-from Basic Aspects to Laboratory Tools, 1*.

Vemuri, P. K., Sagi, H., Dasari, S., Prabhala, A., et al. (2016). Comparative study on various methods for recovery of deoxyribonucleic acid from agarose gels to explore molecular diversity. *Asian Journal of Pharmaceutical and Clinical Research, 9*, 422–424.

Yu, Z., & Morrison, M. (2004). Improved extraction of PCR-quality community DNA from digesta and fecal samples. *BioTechniques, 36*(5), 808–812.

Zou, Y., Mason, M. G., Wang, Y., Wee, E., Turni, C., Blackall, P. J., & Botella, J. R. (2017). Nucleic acid purification from plants, animals, and microbes in under 30 seconds. *PLoS Biology, 15*(11), e2003916.

# Part III
# Major Applications of DNA and RNA

# Transformation or Genetic Modification of Cells/Organisms

**Abstract** Genetic engineering technology has been employed for a long time to enhance the quantity and quality of a product and produce the desired product from an organism. As a critical biotechnological tool, genetic engineering and modifications have solved problems like food security, biofuel production, etc. Additionally, these methods have also been employed to design novel biological systems by modifying the existing ones by either introducing transgenes or reconstructing the present genetic makeup. This chapter describes the different methods like nuclear transformation, biolistic transformation, electroporation, sonoporation, etc., which have been employed to manipulate the genetic constitution of organisms and applications of these methods.

**Keywords** Genetic engineering · Nuclear transformation · Biolistic transformation · Conjugation · Transduction

## 1 Introduction

The history of modifying different organisms dates back to as early as 10,000 years ago in southwest Asia (Raman, 2017). With the establishment of human civilizations, the domestication of different organisms started, and these modifications to produce desired traits along with evolutionary processes resulted in organisms different from their ancestors. The discovery of DNA as genetic material and elucidation of its complex structure led to the beginning of an era of DNA-related science and technology. With the advancement in tools and technology today, biotechnology has become so precise in the sense that it allows biologists to bring genetic modification in an organism up to the level of a single gene. Genetic modification is a discipline under biotechnology that aids biologists in manipulating the genetic structure of organisms using different techniques like simple selection, interspecies

© The Author(s), under exclusive license to Springer Nature Switzerland AG 2022
A. Gautam, *DNA and RNA Isolation Techniques for Non-Experts*, Techniques in Life Science and Biomedicine for the Non-Expert,
https://doi.org/10.1007/978-3-030-94230-4_18

crossing, somaclonal variation, and different genetic engineering tools like microbial vectors, microprojectile bombardment, etc. (Health, 2004a, 2004b). These modifications facilitate the betterment of the organism in terms of quality and quantity or sometimes both. However, the importance of genetic modification is not limited only to the amelioration of organisms. It is an indispensable tool in synthetic biology, where it helps design new biological systems or remodel existing ones for useful purposes (El Karoui et al., 2019). The methods of genetic modification can be broadly divided into two types: non-genetic engineering methods and genetic engineering methods (Table 1).

Genetic modification is an essential discipline of biotechnology that aided scientists in bringing changes in the genetic makeup of different organisms for valuable purposes. Genetically modified microbes are being harnessed for the production of beneficial enzymes and in food industries. Crop improvement using genetic engineering provides resistance against pathogens and leads to increased yield and nutritional value and herbicidal resistance (Datta, 2013; Schütte et al., 2017).

**Table 1** Various methods of genetic modification in plants

| Non-genetic engineering methods | **Simple selection**: Selection of "superior" individuals from the population and its continuous propagation |
|---|---|
| | **Embryo rescue**: Rescuing the embryo from naturally pollinated plants that are not viable in vivo and growing them in vitro conditions |
| | **Somatic hybridization (cell fusion)**: Fusion of protoplast from two different cells, resulting in combined genetic material |
| | **Somaclonal variation**: This refers to the spontaneous mutations that occur among in vitro cultures<br>**Crossing**: Obtaining a hybrid by artificially mating two different sexually compatible varieties |
| | **Interspecies crossing**: Sometimes, by natural means or by human intervention, two closely related species may cross-pollinate and lead to the exchange of genetic information leading to speciation |
| | **Mutation breeding**: It is the method of exposing plants or seeds to mutagenic agents like UV or chemicals such as EMS (ethyl methanesulfonate) for bringing genetic modification |
| | **Cell selection**: This includes isolating the cell population from "elite plants" having desired agricultural traits and culturing those cells. Cells with the phenotype are selected and developed into plants |
| Genetic engineering methods | **Microbial vectors**: This technique exploits the ability of *agrobacterium* sp. to transfer a part of its DNA (called T-DNA) in plants. The gene of interest (GOI) is inserted into the T-DNA region, and after infection, GOI is transferred to the plant |
| | **Microprojectile bombardment**: In this method, the DNA is delivered into plants by adhering them to particles like gold and tungsten and then bombarding them with high pressure |
| | **Electroporation**: In this method, the cells take up genetic material from their surrounding. Exposing organisms to electric impulse destabilizes their cell wall temporarily, allowing DNA to enter inside. After some time, cell walls get reformed. This method is used for transforming microscopic organisms like yeast, algae, etc. |

## 2 Natural Method of Genetic Modification

**Conjugation**: For the first time in 1946, Joshua Lederberg and Edward Tatum, for the first time, gave experimental proof of genetic recombination in bacteria. In one of their experiments on *E. coli*, they discovered that genetic material was being exchanged between two strains of different strains. This recombination led to the exchange of genes between two different strains, which conferred the ability to survive on media on which they were unable to sustain before. This process is called conjugation. However, for successful genetic recombination, there should be physical contact (as proven in the U-shaped tube experiment of Bernard Davis) between two different strains, as separation filters in the U-tube experiment led to no recombinants. The transfer of genetic material is controlled by the fertility factor (F factor). The strain from which genetic material is transferred to another strain is called $F^+$ strain, while receiving strain is called $F^-$ strain. The fertility factor makes this conjugation process unidirectional, and thus it's not genuinely sexual reproduction as in this process, genetic material from both partners is exchanged.

**Transduction**: In nature, genetic exchange in bacteria is sometimes facilitated by a class of viruses called bacteriophages. In 1951, during the recombination experiment in *Salmonella typhimurium*, Joshua Lederberg and Norton Zinder discovered that recombinants were discovered from U-tube when the pore size of filters was varied. Since filters were present between two arms of the U-tube, the presence of recombinants due to conjugation was ruled out. However, later on, it was discovered that recombinants were formed due to virus p22 (salmonella virus), a type of bacteriophage. This type of process of gene transfer is called transduction. During the lytic phase of the phage life cycle, phage somehow picks bacterial genes integrated into the host bacterial chromosome during lysogenic phage, leading to the exchange of genetic material.

## 3 Genetic Modification in Plants

With the population boom, as the global demand for food is increasing, it is leading to crop improvement. Genetic modification of crops has increased their productivity and reduced production cost, enhanced nutrient composition, and food quality. The crops have also been engineered for tolerating different types of biotic and abiotic stresses, allowing them to grow in conditions where they might not otherwise flourish (Phillips, 2008). Various methods, ranging from conventional artificial selection to modern genetic engineering, have been employed to modify plants to produce the desired traits in them.

Of all these methods mentioned above, the most common method for genetic modification is *Agrobacterium*-mediated transformation in plants. This method is a powerful tool for delivering genes of interest into plant cells (Hwang et al., 2017; Jones et al., 2005) and has been discussed for lettuce in brief here:

(a) <u>Growth and preparation of donor plant</u>: Sterile the seeds of lettuce (*Lactuca sativa*) by rinsing it for 1 min with 1.2% sodium hypochlorite, and then wash it with double-distilled water. Now, rinse it for 1 min with 70% ethanol and again give final wash with double-distilled water for 2 min. Spread the seeds on ½ MS medium and keep the plates in the 25 ± 2 °C. After germination, pre-culture 7 days old cotyledons upside down (abaxial) for 3 days on MS medium plate containing plant growth regulators (PGR) for active cell division.

(b) <u>Preparation of *Agrobacterium* cells for infection</u>: Inoculate *Agrobacterium tumefaciens* (strain EH105)-carrying vector with the gene of interest into 50 mL YEP medium containing yeast extract (10 g/L, bacterial peptone 10 g/L, NaCl 5 g/L), rifampicin (10 mg/L), and kanamycin (50 mg/L). Incubate the culture at 28 °C overnight on an orbital shaker at 120 rpm. Centrifuge the bacterial culture ($OD_{600}$ 0.6) at 3000 rpm for 10 min, and then resuspend it in infection medium ($MgCl_2$, 10 mM; MES (2-(N-morpholino) ethanesulfonic acid), 10 mM; acetosyringone, 150 µM).

(c) <u>Inoculation of *Agrobacterium tumefaciens* with *Lactuca sativa* explants</u>: Incubate the infection medium at room temperature for half hour. Co-cultivate the pre-cultured explants with infection medium for 10 min. Dry the explants by keeping them on autoclaved blotting paper. Place the dried explants upside down on MS media containing PGR and hygromycin. Incubate the plants in the dark at 25 °C for 3 days. After 3 days, incubate them at 25 °C on a selection plate (containing PGR and hygromycin) and subculture after 14 days.

(d) <u>Shoot regeneration and transgenic confirmation by PCR:</u> Shoot regeneration is generally seen from the callus after 28–30 days. Isolate DNA from the plant and perform PCR for confirming the transgenics.

(e) <u>Rooting in transgenic plants and shifting to the greenhouse:</u> Keep the shoots on ½ MS medium for root formation. After the formation of roots, the plants are shifted to the greenhouse for further growth.

Another suitable and efficient method of genetic modification in the plant is by delivering the DNA in plant cells by particle bombardment. The particle bombardment method has also been used to produce desired products for organelle-specific transformation like chloroplast transformation (Klein & Fitzpatrick-Mcelligott, 1993; Kumar et al., 2004). The protocol must be optimized with different organisms. The amount of DNA per shot, distance, and shot pressure should be optimized (Fig. 1). The general protocol for particle bombardment using is described below.

<u>Basic steps before bombardment pressure of leaf tissues</u>: Keep all suspensions and buffers on ice in the laminar airflow. Sonicate gold particle suspension before use to disassociate the aggregated particles. Sterilize all the instruments, forceps, etc. before use. Wipe gene gun with 70% ethanol before use. Sterilize rupture discs and macrocarriers by immersing in 70% ethanol. Keep the pressure in regulator 200 psi above by helium gas to burst the rupture discs at selected pressure (e.g., set pressure 1300 psi when using the 1100 psi ruptured disc).

<u>Steps for coating DNA on gold particles</u>: Dilute the gold carrier particles (DNAdelTM gold carrier particles S550d, Seashell technology) into the binding

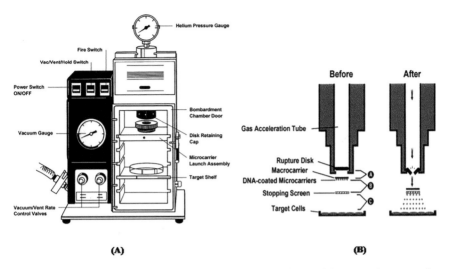

**Fig. 1** (**a**) Biolistic PDS-1000/He gene gun setup (**b**) events in disc retaining cap and macrocarrier launch chamber before and after bombardment

buffer provided with the kit. For dilution, suspend 1.5 mg gold (30 μL volume of 50 mg/mL gold particle suspension) into 20 μL binding buffer in a 1.5 mL microfuge tube. Add DNA to the gold particle at the ratio of 2–5 μg DNA per mg gold (calculation should be done according to DNA for each bombardment). Vortex the mixture briefly. Add precipitation buffer in equal amount and keep on ice for 3 min. Centrifuge the whole mixture at 10,000 rpm in a microfuge (Eppendorf) for 30 seconds to pellet down the DNA-coated gold particles. Discard the supernatant and add 500 μL of absolute ice-cold ethanol, vortex briefly and centrifuge at 10,000 rpm in a microfuge (Eppendorf) for 30 s to wash the pellet. Remove the supernatant and add ice-cold absolute ethanol to desired gold concentration. Sonicate briefly. Apply the mixture on the macrocarrier, let it dry, and then place it on the macrocarrier holder with the help of a key. Place rupture disc of desired pressure in disc retention cap.

Bombardment of leaf tissue with desired DNA or vector coated on gold particles: Place the leaf upside down (abaxial surface up and adaxial surface down; Fig. 2) on MS medium, and place it on the target shelf in the chamber of Biolistic PDS-1000/He gene gun (Fig. 1a, b) and close the door. Evacuate the chamber and hold the vacuum at the desired level (minimum 5 inches of mercury level). Press the fire button until the rupture disc bursts and the helium pressure gauge drops to zero. After this process is over, release the fire button. After firing is done, release helium pressure from the system.

Process after the bombardment of leaf tissue: Keep the bombarded leaf for 2 days in the dark in the culture room on a selection-free MS medium containing PGR. On the third day, dissect the bombarded leaves in small explant (approx. 5 × 5 mm square meter size) and inoculate on selection MS medium containing selection antibiotics and PGR, and keep in light for shoot induction for 3–4 weeks. Maintain

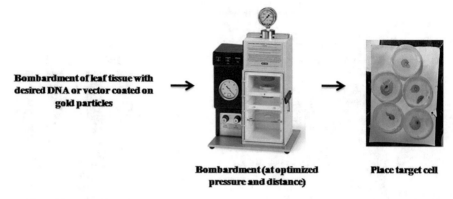

**Bombardment of leaf tissue with desired DNA or vector coated on gold particles**

**Bombardment (at optimized pressure and distance)**

**Place target cell**

**Fig. 2** Biolistic transformation in lettuce plant. (Image courtesy: Metabolic engineering group, ICGEB (New Delhi))

explants on selection medium by subculturing until shoots are induced. Transfer transgenic shoots to rooting medium. After shoot formation, the plants are shifted to the greenhouse for further growth.

Various new plant genetic modification techniques like CRISPR, RNAi, and many more have been developed with advancements in tools and technology. These techniques, along with modifications in the old one, make genetic modifications in plants more precise and stable.

## 4 Genetic Modification in Animals and Microbes

Biotechnology has empowered scientists to manipulate not only the genetic structure of plants but also complex animals and microbes. Modification in animals, especially cattle, for better productivity is not a new idea. However, assisted reproductive technique (ART) and molecular biology gave new dimensions to gene transfer methods in mammals (Bihon Asfaw & Assefa, 2019). Some of the commonly used methods for genetic modification in animals are given in Table 2. However, the modification methods are not limited to this. There are other methods of genetic modifications like recombinant DNA technology (e.g., GloFish creation), DNA homologous recombination, and DNA delivery by viral vectors. Many techniques are employed for modification in microbe genetic structure, and such microbes are known as GEM (genetically engineered microbes) (Hanlon & Sewalt, 2021).

Some of the methods for modifying the genetic makeup of microbes are as follows.

**Table 2**  Methods of genetic modification in animals

| Pronuclear microinjection | This method includes the transfer of exogenous DNA to pronuclei of the fertilized egg with the help of microinjection. Microinjected eggs are transplanted into pseudopregnant foster mother oviducts, where they develop into viable individuals (Liu et al., 2013; Rülicke & Hübscher, 2000) |
|---|---|
| ES cell microinjection | In this method, the undifferentiated embryonic stem (ES) cells are first isolated from the inner mass of blastocysts of donor animals. After culturing of ES cells, transgenes are inserted into it by various methods. After that, the ES with transgene is injected into the blastocyst, further implanted into the pseudopregnant host (Du et al., 2019; ES Cell Microinjection (Mice), 2019) |
| Somatic cell nuclear transfer | In this method, the nucleus of a somatic cell is transferred to the enucleated oocyte. This leads to the generation of a genetically identical organism to somatic cell donors (Matoba & Zhang, 2018; Tian et al., 2003) |

# 5  Non-targeted Mutagenesis

It is one of the oldest methods for genetically modifying microorganisms. In this method, the microbes are subjected to chemical mutagens or UV radiation. Since this technique leads to random mutations, the final genetic makeup remains unknown, and only by high-throughput DNA sequencing the final genetic structure can be revealed.

# 6  Genetic Engineering

With the help of molecular biology and recombinant DNA technology, scientists today can produce robust and productive hosts that are not pathogenic and nontoxic for use. Once a suitable host is identified, by using molecular techniques, the gene sequences can be inserted in or deleted from the genome of the microbe. The expressed gene can be either endogenous or exogenous. In genetic engineering, gene cloning is an important aspect. The selection of a suitable vector is a prerequisite for gene cloning as it is a pivotal tool. A suitable vector should be of low molecular weight, have the ability to confer readily selectable phenotypic traits on host cells, possess MCS (multiple cloning site), and have the capacity to replicate within host cells to produce multiple copies. Vectors can be of a different type, namely, plasmid, λ phage, P1 phage, and artificial chromosomes (YACs, yeast artificial chromosomes; BACs, bacterial artificial chromosomes; and HAC, human artificial chromosome) (Fig. 3). These vectors have different possible insert size which is summarized in Table 3.

A cloning vector is a DNA fragment having the capability to self-replicate inside the host organism. The cloning vector must have an ori (origin of replication), restriction site, antibiotic resistance gene, multiple cloning sites, and compatibility with the host. Plasmids (pBR322, F plasmid, pUC18, Col plasmid, etc.) can encode only self-replicated proteins. Phage λ and M13 phage are the bacteriophages which

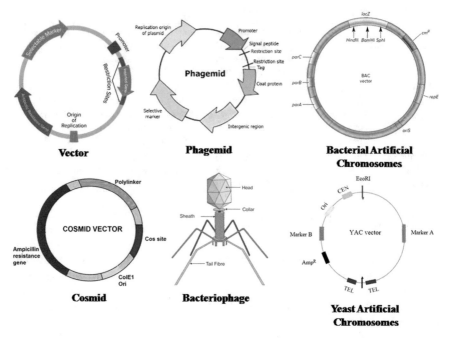

**Fig. 3** Cloning vectors and their types

**Table 3** Vectors and their possible insert size in respective hosts

| S. No. | Vector | Host | Insert size |
|--------|--------|------|-------------|
| 1 | Plasmid | *E. coli* | 5–25 KB |
| 2 | λ phage | *E. coli* | 35–45 KB |
| 3 | P1 phage | *E. coli* | 110–115 KB |
| 4 | Phage artificial chromosome (PAC) | *E. coli* | 100–300 KB |
| 5 | Bacterial artificial chromosome (BAC) | *E. coli* | <300 KB |
| 6 | Yeast artificial chromosome (YAC) | *Saccharomyces cerevisiae* | 200–2000 KB |
| 7 | Human artificial chromosome (HAC) | Cultured human cells | >2000 KB |

are mainly utilized in gene cloning. Cosmids are another type of vector having the capability to incorporate with λ DNA segment of bacteriophage. The artificial vectors such as Phagemids (which have F1 ori, inducible lac gene promoter and are used with M13 phage and HACs) are utilized for gene delivery or gene transfer into human cells. BACs show similarity with *E. coli* plasmid and encode ori binding proteins, and YACs are generally used for cloning inside eukaryotic cells and contain yeast telomere at each end. Some retroviral vectors are used to introduce novel or manipulated genes into animal or human cells.

**Fig. 4** Genetic
transformation through
sonoporation

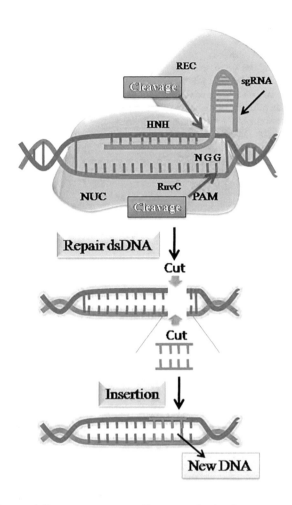

There are many nonviral gene delivery systems too. For example, in electroporation, the high-voltage electric current creates the temporary pores in the cell membranes for the transfer of DNA, and in sonoporation, ultrasound waves form the
pores in the cell membrane for the transfer of nucleic acid materials (Fig. 4). Similarly,
the transfer of DNA by liposomes or lipofection is considered a safer and less toxic
method as compared to the viral vectors. CRISPR is one of the most recent tools used
for editing the genome of microbes. It allows scientists to insert desired DNA
sequences at the exact spot (Jinek et al., 2012) (Fig. 5).

**Fig. 5** In the CRISPR/
Cas9 system, a guide RNA
(sgRNA) instantly
hybridizes a 20-nucleotide
DNA sequence, resulting
in a double-strand break.
The broken DNA works as
substrates for endogenous
cellular DNA repair
machinery catalyze the
nonhomologous end
joining or homology-
directed repair

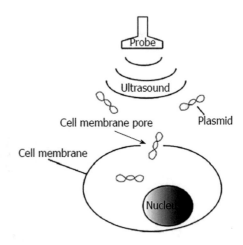

# References

Bihon Asfaw, A., & Assefa, A. (2019). Animal transgenesis technology: A review. *Cogent Food & Agriculture, 5*(1), 1686802.

Datta, A. (2013). Genetic engineering for improving quality and productivity of crops. *Agriculture & Food Security, 2*(1), 1–3.

Du, Y., Xie, W., Zhang, F., & Liu, C. (2019). *Chimeric mouse generation by ES cell blastocyst microinjection and uterine transfer. In microinjection* (pp. 99–114). Humana Press.

El Karoui, M., Hoyos-Flight, M., & Fletcher, L. (2019). Future trends in synthetic biology-a report. *Frontiers in Bioengineering and Biotechnology, 7*, 175.

ES Cell Microinjection (Mice). (2019). The University of Edinburgh. https://www.ed.ac.uk/bioresearch-veterinary-services/facilities/transgenics/transgenic-mouse-production/es-cell-injection.

Hanlon, P., & Sewalt, V. (2021). GEMs: Genetically engineered microorganisms and the regulatory oversight of their uses in modern food production. *Critical Reviews in Food Science and Nutrition, 61*(6), 959–970.

Health, N. R. C. (US) C. on I. and A. U. E. of G. E. F. on H. (2004a). Introduction. In *Safety of genetically engineered foods: Approaches to assessing unintended health effects*. National Academies Press (US). https://www.ncbi.nlm.nih.gov/books/NBK215770

Health, N. R. C. (US) C. on I. and A. U. E. of G. E. F. on H. (2004b). Methods and mechanisms for genetic manipulation of plants, animals, and microorganisms. In *Safety of genetically engineered foods: Approaches to assessing unintended health effects*. National Academies Press (US). https://www.ncbi.nlm.nih.gov/books/NBK215771

Hwang, H. H., Yu, M., & Lai, E. M. (2017). Agrobacterium-mediated plant transformation: Biology and applications. *Arab Book, 15*, e0186.

Jinek, M., Chylinski, K., Fonfara, I., Hauer, M., Doudna, J. A., & Charpentier, E. (2012). A programmable dual-RNA–guided DNA endonuclease in adaptive bacterial immunity. *Science, 337*(6096), 816–821.

Jones, H. D., Doherty, A., & Wu, H. (2005). Review of methodologies and a protocol for the agrobacterium-mediated transformation of wheat. *Plant Methods, 1*(1), 1–9.

Klein, T. M., & Fitzpatrick-Mcelligott, S. (1993). Particle bombardment: A universal approach for gene transfer to cells and tissues. *Current Opinion in Biotechnology, 4*(5), 583–590.

Kumar, S., Dhingra, A., & Daniell, H. (2004). Manipulation of gene expression facilitates cotton plastid transformation of cotton by somatic embryogenesis and maternal inheritance of transgenes. *Plant Molecular Biology, 56*(2), 203–216.

Liu, C., Xie, W., Gui, C., & Du, Y. (2013). Pronuclear microinjection and oviduct transfer procedures for transgenic mouse production. In *Lipoproteins and cardiovascular disease* (pp. 217–232). Humana Press.

Matoba, S., & Zhang, Y. (2018). Somatic cell nuclear transfer reprogramming: Mechanisms and applications. *Cell Stem Cell, 23*(4), 471–485.

Phillips, T. (2008). Genetically modified organisms (GMOs): Transgenic crops and recombinant DNA technology. *Nature Education, 1*(1), 213.

Raman, R. (2017). The impact of genetically modified (GM) crops in modern agriculture: A review. *GM Crops & Food, 8*(4), 195–208.

Rülicke, T., & Hübscher, U. (2000). Germline transformation of mammals by pronuclear microinjection. *Experimental Physiology, 85*(6), 589–601.

Schütte, G., Eckerstorfer, M., Rastelli, V., Reichenbecher, W., Restrepo-Vassalli, S., Ruohonen-Lehto, M., Saucy, A. G. W., & Mertens, M. (2017). Herbicide resistance and biodiversity: Agronomic and environmental aspects of genetically modified herbicide-resistant plants. *Environmental Sciences Europe, 29*(1), 1–12.

Tian, X. C., Kubota, C., Enright, B., & Yang, X. (2003). Cloning animals by somatic cell nuclear transfer–biological factors. *Reproductive Biology and Endocrinology, 1*(1), 1–7.

# Gene Cloning and Vectors

**Abstract** "Gene (or DNA) cloning" is the process of generating identical copies of a cell or an organism. Clones thus produced possess identical genes. In this technique, restriction enzymes cut DNA carrying the target gene(s) into fragments. These segments are subsequently cloned onto a cloning vector, resulting in a recombinant DNA molecule that is then transmitted to the appropriate host cell. The basic principle is to amplify the target gene within the specific cell using its machinery, which can be used for further downstream applications, including sequencing, genotyping, and heterologous expression of a protein. Gene cloning is similar to polymerase chain reaction (PCR) in this way, albeit there are some significant distinctions between the two procedures. It is feasible to preserve a duplicate of any specific segment of DNA and make a limitless amount of it through cloning. Cloning is a genetic engineering technique with broad applications in medicine. Researchers can produce many specific proteins of potential therapeutic values, use transformed DNA to correct some human genetic defects, and synthesize new vaccines. Cloning has wide applications in agriculture and farm for improving farm animals and crop plants for better production and introducing new functions into plants and animals. The limitation of this technique lies in the fact that mutation during cloning may result in defective protein production.

**Keywords** Applications · Cloning · Expression · Gene · Vector · Protein

## 1 Introduction

The word "clone" comes from the ancient Greek word *kln* (which means twig), which alludes to the act of creating a new plant from a twig. Herbert J. Webber coined this term. The terms "gene cloning," "DNA cloning," "molecular cloning," and "recombinant DNA technology" all refer to the same technique. In simple words, "gene cloning" is the process of generating an identical copy of a cell or an organism. Clones represent organisms that share identical genes. In scientific terms,

© The Author(s), under exclusive license to Springer Nature Switzerland AG 2022      147
A. Gautam, *DNA and RNA Isolation Techniques for Non-Experts*, Techniques in
Life Science and Biomedicine for the Non-Expert,
https://doi.org/10.1007/978-3-030-94230-4_19

gene or DNA cloning is a technique of genetic engineering whereby foreign DNA (or gene) can be introduced into a host (bacterial, plant, or animal) cell (Brown, 2020).

Restriction enzymes are used to fragment the DNA containing the target gene(s) into pieces. These fragments are then inserted into a cloning vector to form a recombinant DNA molecule and then transferred to the suitable host cell, such as the bacterium *Escherichia coli*. Within the host cell, the recombinant DNA molecule undergoes replication, and the host cell produces progeny, all of which contain the inserted gene. These identical cells produce colonies that are called "clones" (Fig. 1). The gene carried by the recombinant DNA molecule is now said to be cloned because each cell in the clone contains one or more copies of it (Lodge et al., 2007).

The cloned DNA can subsequently be studied or the gene-encoded protein can be produced. We may want to transfer the cloned DNA into another organism for many applications, but the initial cloning steps are almost always performed in *E. coli*. This technique provides researchers an opportunity to study the structure of genes of interest in detail.

Based on the requirement of a host cell, there are two types of gene cloning techniques: (a) cell-based DNA cloning and (b) cell-free DNA cloning (polymerase chain reaction, PCR).

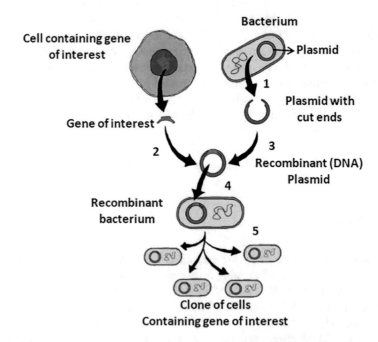

**Fig. 1** Showing steps of gene cloning. (*1*): Isolation of the plasmid and gene of interest (*2*): Combining gene of interest with a plasmid (*3*): Recombinant plasmid (DNA) formation (*4*): Transfer of recombinant plasmid into bacterium (vector) (*5*): Multiplication of recombinant bacterium forming clones or duplicate colonies

## 2  Basic Steps of the Technique (Cell-Based DNA Cloning)

**1. Identification and isolation of the desired gene**: The target gene is identified in the organism's DNA. The required DNA is isolated from the organism, purified, and fragmented using restriction enzymes or restriction endonucleases (RE). These enzymes were called so because they are responsible for restrictions imposed by the host. These enzymes are produced by the bacteria that cut the DNA at specific sites and give fragments with cohesive ends. Each fragment consists of a single-stranded sequence of nucleotides at its ends that will form hydrogen bonds with the complementary base pairs on the single-stranded stretches of other DNA fragments using enzyme ligase. A restriction site is a DNA sequence that these enzymes recognize, and restriction fragments are DNA fragments produced by cutting with these enzymes (Rapley, 2000).

**2. Insertion of DNA fragment into cloning vector**: The desired DNA fragment containing the gene of interest to be cloned is inserted into a cloning vector to produce a recombinant DNA molecule. A cloning vector is a small circular molecule of DNA found in many types of bacteria, into which a foreign DNA fragment can be inserted. Plasmids are the most commonly used vectors for gene cloning. The type of plasmid used in the gene cloning has a single restriction site and, when cleaved by the same RE, gives rise to cohesive ends complementary to the DNA of interest. The enzyme ligase then forms the phosphodiester bond.

**3. Transformation of the vector into the host cell**: The cloning vector with the desired gene (i.e., recombinant vector) is inserted into the host cell, mostly a bacterial cell (most common being *E. coli*). Transformation can be natural where bacteria take up recombinant vectors automatically. For example, *Bacillus*, *Haemophilus*, *Helicobacter pylori*, etc. are naturally competent. Transformation can be by chemical methods where the host cell is kept in calcium chloride, which enables the host cell to take up vector, or by electroporation, where pores appear in the host cell membrane through which vectors can enter inside the cell.

**4. Initiation of cloning**: Once the transformation is over, recombinant host cells must be separated from the bacterium by introducing them in the fresh culture media. For proper growth and multiplication, optimum parameters are necessary at this stage. The host cells undergo division and redivision along with a replication of the recombinant DNA carried by them. Depending upon the aim of cloning, numerous copies only of the gene of interest may be required, then only replication of the host cell is allowed. On the other hand, the gene of interest can be transcribed and translated into proteins when required. The gene is expressed, with the gene product being a protein. The process is called "expression" (Lodge et al., 2007).

**5. Isolation and purification**: Isolation of the gene of interest with the vector or the protein encoded by it is the final step in the technique of gene cloning. Following that, the isolated gene copy/protein is purified.

On the basis of these steps, we can say that the following four items are required for gene cloning:

a. Fragment of DNA with the gene of interest
b. Restriction enzymes/endonucleases and ligase enzymes
c. Cloning vectors—to carry, maintain, and replicate cloned gene in the host cell
d. Host cell—in which recombinant DNA can replicate

## 3  Vector or Cloning Vector

A cloning vector is defined as a small segment of DNA into which a foreign DNA fragment or DNA of choice is to be incorporated for the process of cloning. The type of host cells and the goals of the cloning experiment dictate the vector to be utilized. The study of the molecular structure of DNA has been assisted by these vectors. Any molecule of DNA that has to be cloned must first be integrated into a cloning vector. These DNA elements can be replicated and maintained in a host organism if the vector has replication capabilities. The bacterium is a typical host organism that is a fast-growing and replicating bacterium. As a result, any vector with an *E. coli* replication origin will efficiently replicate (including any integrated DNA). As a result, any DNA cloned into a vector will facilitate the multiplication of the inserted foreign DNA fragment, as well as subsequent evaluation can be undertaken. The cloning method is similar to polymerase chain reaction (PCR) in this way, albeit there are some significant distinctions between the two procedures. It is feasible to not only preserve a duplicate of any specific segment of DNA but also to make a limitless amount of it through cloning.

A cloning vector should possess the following features:

1. Ori (origin of replication) is the classical feature of a cloning vector, which represents a specific sequence of nucleotides, which severs as the starting point of replication.
2. It must contain a restriction site into which the target DNA can be introduced.
3. Its size should be small (less than 10 kb) to easily enter the host cell, as large molecules are more difficult to manage and prone to breakdown during purification.
4. The vector should possess multiple cloning sites (MCS).
5. In order to facilitate screening of the recombinant organism and to allow the host to grow on specified conditions, a selectable marker gene with an antibiotic resistance gene should be present in the vector; it can selectively amplify this specific vector in the host cell.
6. It should be able to work both in the eukaryotic and prokaryotic environments.

## 4  Types of Cloning Vectors

Many cloning vectors have been discovered, mostly based on naturally existing molecules like bacterial plasmids or bacteriophages or combinations of the elements that make them up, such as cosmids. The choice for the type of vector to be used for the construction of the gene library is usually connected to the ease of manipulations required, and the maximal size of the vector's foreign DNA insert. Broadly, vectors have been classified into two types (Fig. 2):

a. **Cloning vector**—It is designed to carry the gene of interest into the host cell. Its main function is to increase the copy number of the gene of interest within the host cell. This is possible by the presence of the origin of replication within the

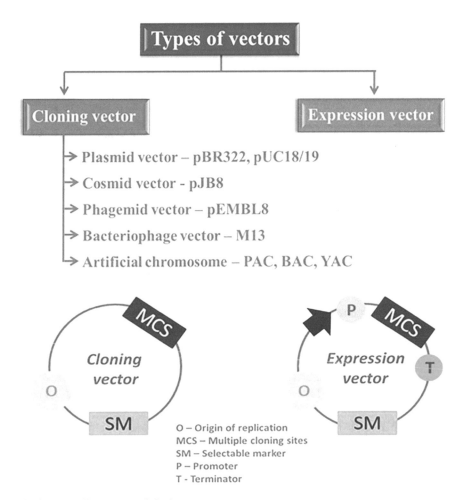

**Fig. 2** Two different types of cloning vectors

cloning vector. The host cell carrying a cloning vector can be recognized or selected with the help of a selectable marker (Brown, 2020).
b. **Expression vector**—This type of vector carries a dual function, i.e., carries the foreign gene; it's the host cell as well as their expression within the host cell. Hence, it contains sequences required for replication (origin of replication), transcription (promoter, enhancer), translation (ribosome binding site, start and stop codon), selectable marker, and restriction enzyme sites (Brown, 2020).

A brief description of commonly used vectors has been given below:

**1. Plasmid vectors**: Plasmids are the most extensively used vectors for gene cloning that lead to independent existence in bacterial cells. They are extrachromosomal, circular, self-replicating, double-stranded DNA molecules. Plasmid DNA replicates in a manner that is not always synchronous with chromosomal division. Typically, the size of the plasmid vectors that are used for cloning is less than 5 kb. It is difficult to handle large DNA molecules as they may undergo degradation. One or more genes are almost always present in plasmids which typically are responsible for the host bacterium's valuable trait.

Many structural elements are present in a plasmid to be used for cloning (Wong, 2006). Plasmids represent an "origin of replication" that directs plasmid replication and ensures that when the cell divides, numerous copies of the plasmid are disseminated throughout the daughter cells. Plasmids also contain multiple cloning sites (MCS) that artificially produced recognition sequences for various restriction enzymes. It intends to make the insertion of foreign DNA easier. The promoter region upstream of the multiple cloning site is present in some plasmids and called expression vectors. Transcription and translation of the inserted DNA are possible due to this construction. Artificial plasmids are created on a regular basis. pRB322, for example, has a replication origin, 2 antibiotic resistance genes, and approximately 40 distinct restriction sites. The size of the DNA that can be introduced limits the use of plasmids (Wong, 2006; Clark & Pazdernik, 2013).

**pBR322**—It is the most commonly used cloning vector nowadays. Bolivar and Rodriguez synthesized it in 1977 at the University of California, hence named pBR322, where p stands for plasmid and 322 is the serial number. Its size is 4363 base pair. It contains the rep gene, which is required for replication of the plasmid. Additionally, it contains the "tet" and "bla" genes which encode tetracycline resistance and β-lactamase, providing resistance to tetracycline and ampicillin, respectively (Clark & Pazdernik, 2013).

**pUC18/19**—This vector is synthesized by Messing et al. at the University of California, hence named pUC. Its size is 2686 bp. It is present in very high copy numbers, hence ideal for cloning. It contains pMB1 as the origin of replication. Additionally, it contains multiple cloning sites within the lacZ gene, which are accessible to a large number of restriction endonucleases and allow the rapid visual detection of recombinant clones by blue-white screening. It also contains a bla gene that encodes the β-lactamase enzyme and provides resistance against ampicillin antibiotics (Rodriguez & Denhardt, 2014).

Some of the important genes present in plasmids include:

**Fertility plasmid**: This confers the conjugation between bacteria, allowing gene transfer from one generation to another.

**Resistance (R) plasmid**: This enables the bacterium to be resistant to antibiotics. Some Col factors and R factors can move from one cell to another, allowing them to spread rapidly throughout a microbial population.

**Colicinogenic (Col) plasmid**: Factors that regulate the formation of antibiotic-active proteins termed colicins that have an antibiotic activity that codes the protein named bacteriocin, which helps in the killing of other bacteria.

**Degradative plasmid**: This allows the host bacterium to break down unusual/unwanted/harmful molecules such as toluene and salicylic acid.

**Virulence plasmid**: This renders the host bacterium pathogenic; for example, these include the Ti plasmids of *Agrobacterium tumefaciens*, which induce crown gall disease on dicotyledonous plants.

**2. Bacteriophage vectors**: Viruses that infect bacteria specifically and inject phage DNA into the host cell, where it replicates, are known as phages. Phages possess a very simple structure, consisting of a DNA (or occasionally RNA) molecule containing many genes, including many for phage replication, wrapped by a protein-based protective shell or capsid. They infect the bacteria in two steps (known as phage infection cycle): (i) attachment of the phage particle to the bacterial cell wall followed by inoculation of its DNA into the cell and (ii) replication of the phage DNA, usually by specific phage enzymes coded by genes in the phage chromosome. The entire infection cycle is completed relatively quickly with some phage species, often less than 20 min. Such type of infection cycle is known as the lytic cycle, as the lysis of the bacterial cell is coupled with the release of additional phage particles.

**Cosmid**—Collins and Hohn firstly described the cosmid vector in 1978. Cosmid is the hybrid vector containing sequences of bacterial plasmid and bacteriophage λ. Cosmid stands for cos and mid, where cos is a sequence derived from lambda phage and mid is from the word plasmid. The cos sequence helps in the circularization of the plasmid once it enters the host cell. Cosmid contains ColE1, a bacterial origin of replication that helps cosmid to replicate in bacteria. Additionally, it contains ampicillin and tetracycline resistance genes which help in the selection of transformed cells. It also contains sites for restriction endonucleases that aid in the ligation of DNA fragments during cloning. Cosmid vector can accommodate insert size up to 45 kb. An example includes pJB8 (Clos & Zander-Dinse, 2019).

**Phagemid**—It is a type of cloning vector that possesses features of both the plasmids and bacteriophages. It contains an origin of replication from a plasmid as well as from a bacteriophage. It is derived from the pUC19 plasmid vector having the ColE1 origin of replication. It also contains the lacZ gene from Lac promoter in multiple cloning sites, which is accessible to restriction endonucleases and hence helps in cloning. It has an ampicillin-resistant gene that helps in the screening and selection of positive recombinant clones. An example includes the pEMBL series of vectors (Qi et al., 2012).

**M13**—This is a unique bacteriophage vector having single-stranded DNA. Its size is 6.4 kb and was sequenced by Sanger in 1982. This vector infects F+ bacterial cells only and enters the cytoplasm by F-pili, where its single-stranded DNA is converted into double-stranded DNA utilizing host cell machinery (Green & Sambrook, 2017).

**P1-derived artificial chromosome (PAC)**—PAC is a type of phage-based *Escherichia coli*-compatible vector. It is a high cloning capacity vector and can accommodate up to 150 kbp, suitable for cloning the full-length gene. It can be introduced in the host cell by electroporation (Bajpai, 2014).

**Bacterial artificial chromosome (BAC)**—BAC is a unique vector with a cloning capacity of 300 kb; hence, it can be used for cloning the large gene. By this, it has been used in human genome sequencing project. It is easily maintained, and its size is 7.4 kb. It contains oriS and repE genes responsible for replication and regulation of copy numbers. It also contains cloning sites in the lacZ gene, which helps in the blue-white selection of positive recombinant clones. It also contains T7 and Sp6 phage promoters for the transcription of inserted genes (Heintz & Gong, 2020).

**Yeast artificial chromosome (YAC)**—Murray and Szostak had described YAC in 1983. YAC exists as a circular plasmid in *E. coli* and as a linear chromosome in yeast; hence, it contains the origin of replication as well as a selection marker for both *E. coli* and yeast. It can accommodate up to 3000 kb of insert and can be incorporated into the yeast cells by electroporation (Fujiwara et al., 1999).

# 5  Applications

Gene cloning has the following applications (Fig. 3):

1. **Application in medicine:**

   a. One of the significant applications of gene cloning is the production of a huge amount of protein. Many proteins with therapeutic potential can be discovered in trace levels in biological systems. Purification of these proteins from their natural sources is not economically feasible. To avoid this, the gene responsible for specific protein production is inserted into the suitable host system that can efficiently produce the protein in large quantities. This method has been used commonly to produce human insulin, human growth hormone, interferon, hepatitis B vaccine, tissue plasminogen activator, interleukin-2, and erythropoietin (Wong, 2006).

   b. Corrections of some human genetic defects: Genetic disorders caused by deficiency of specific protein or enzyme (single gene defect) such as severe combined immunodeficiency (SCID) can be corrected by introducing a healthy (therapeutic) gene.

   c. Therapeutic vaccines: Recombinant DNA containing the desired gene in plasmids can be used for the synthesis of a few vaccines.

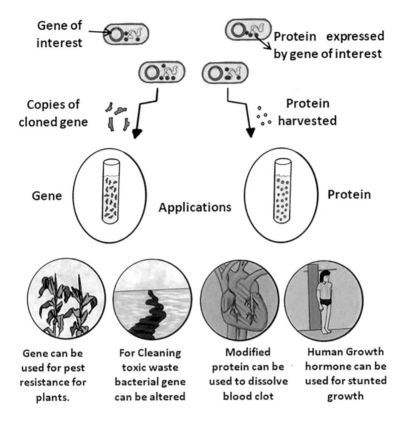

**Fig. 3** Showing applications of gene cloning. Cloning can be of the gene that can be used for pest resistance in plants or for cleaning toxic waste in river water. Cloned protein can be used for making human growth factors or as a clot-dissolving agent

2. **Application in agriculture and farm:**

   a. Improvement of farm animals and crop plants can be made by some alterations like (a) manipulating genes to produce an exogenous (foreign) enzyme or by (b) blocking enzyme production, which remains the main application of gene cloning in the agriculture field. For example, the inhibition of structural polymers' breakdown in cell walls of ripening tomatoes. Here, the gene expression for the enzyme involved in the breakdown process is blocked using the antisense technique. These modified tomatoes can indeed be left to mature on the vine to fully develop their flavor and color. One can also control the ripening by blocking the expression of the enzyme that catalyzes the crucial step in the formation of the ripening hormone, ethylene (Rapley, 2000; Wong, 2006).

   b. Introduction of new functions into plants and animals: New proteins can be produced by inserting recombinant genes into plants and animals. This method has been applied for the production of pest-resistant plants by cloning an endotoxin of bacteria. Similarly, for farm animals, resistance to particular diseases can be introduced by this gene cloning method. High-quality meat can be produced by the introduction of growth hormone genes into the farm animals, which leads to an increase in growth rate and protein-to-fat ratio for lean meat production.

## References

Bajpai, B. (2014). High-capacity vectors. In *Advances in biotechnology* (pp. 1–10). Springer.

Brown, T. A. (2020). *Gene cloning and DNA analysis: An introduction*. John Wiley & Sons.

Clark, D. P., & Pazdernik, N. J. (2013). *Molecular Biology*. Elsevier Science.

Clos, J., & Zander-Dinse, D. (2019). *Cosmid library construction and functional cloning. In Leishmania* (pp. 123–140). Humana Press.

Fujiwara, Y., Miwa, M., Takahashi, R. I., Kodaira, K., Hirabayashi, M., Suzuki, T., & Ueda, M. (1999). High-level expressing YAC vector for transgenic animal bioreactors. *Molecular Reproduction and Development, 52*(4), 414–420.

Green, M. R., & Sambrook, J. (2017). Plating bacteriophage M13. *Cold Spring Harbor Protocols, 2017*(10), pdb-prot093427.

Heintz, N., & Gong, S. (2020). Working with bacterial artificial chromosomes (BACs) and other high-capacity vectors. *Cold Spring Harbor Protocols, 2020*(10), pdb-top097998.

Lodge, J., Lund, P., & Minchin, S. (2007). *Gene cloning*. Taylor & Francis.

Qi, H., Lu, H., Qiu, H. J., Petrenko, V., & Liu, A. (2012). Phagemid vectors for phage display: Properties, characteristics and construction. *Journal of Molecular Biology, 417*(3), 129–143.

Rapley, R. (2000). Recombinant DNA and genetic analysis. *Principles and Techniques of Biochemistry and Molecular Biology*.

Rodriguez, R. L., & Denhardt, D. T. (Eds.). (2014). *Vectors: A survey of molecular cloning vectors and their uses*. Butterworth-Heinemann.

Wong, D. W. (2006). *The ABCs of gene cloning*. Springer.

# Polymerase Chain Reaction (PCR)

**Abstract** Polymerase chain reaction or PCR is a test to detect and amplify DNA (or cDNA) in an organism. In principle, this technique is similar to in vivo transcription and cloning. The isolated DNA and RNA (after converting it into cDNA through reverse transcription) serve as the starting material for any PCR procedure. The three basic steps involved in this process are denaturation of the template (generally ds DNA), annealing of the primers (for specific DNA sequence amplification), and, finally, the extension of the DNA stretch. A combination of different enzymes, buffers, and temperatures in program-controlled equipment (known as PCR machine) helps amplify DNA. The final product (known as amplicon) can be detected at the end through agarose gel electrophoresis or analyzed through different detection methods on a real-time basis. PCR has a broad range of applications, including basic research, medical diagnostics, forensics, and agriculture.

**Keywords** PCR · DNA · cDNA · Primers · Annealing · Taq polymerase · Extension

## 1 Concept of PCR

Polymerase chain reaction (PCR) is a common research application for obtaining similar copies of a DNA fragment. It could be considered as a comparable procedure to the in vivo DNA replication as the product is similar: the production of additional complementary DNA segments. Kary Mullis invented PCR in 1983, for which he earned a Nobel Prize in chemistry in 1993 (Shampo & Kyle, 2002). Since then, PCR has evolved as a standard procedure in molecular biology laboratories to rapidly amplify a small stretch of DNA from a mixture of DNA molecules. One of the reasons for the wide range adoption of the PCR is the simplicity of the reaction setup and ease of the experimental manipulation of steps (Fig. 1). It is also a technique that has replaced the traditional DNA cloning methods in many cases since it fulfills the same function, i.e., generating large amounts of DNA from the limited

## PCR components

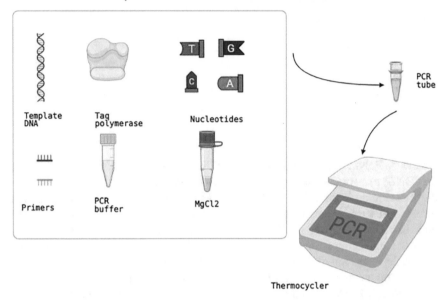

Fig. 1 PCR components and thermocycler

starting material. This takes much less time as compared to DNA cloning. PCR technique is often referred to as "acellular cloning" of a DNA fragment through an automated system, which usually takes several days with standard molecular cloning techniques (Wilson & Walker, 2010). Although PCR seems an effortless technique, it involves knowledge about specific bioinformatics tools and some DNA sequence information. It does require the knowledge of some DNA sequence information that flanks the fragment of DNA to be amplified, often referred to as target DNA. The DNA fragment which acts as a starting material from a complex mixture of starting material usually termed as the template DNA. Three components are critical to designing a PCR-based experiment. These include Taq polymerase, primer pairs, and reaction conditions (Fig. 1). All the required components and their role in PCR have been discussed below:

## 2  Taq Polymerase

Like DNA replication in an organism, PCR requires a DNA polymerase enzyme that makes new strands of DNA, using existing strands as templates. In replication, DNA copies into two daughter DNA molecules by adding dNTPs to the growing DNA. DNA polymerase has dual roles in replication: 5′–3′ exonuclease polymerase activity and 3′–5′ exonuclease proofreading activity. But standard DNA polymerases are unstable at higher temperatures, so they are not suitable for PCR. On the other hand, Taq DNA polymerase plays an essential role in PCR because of its stability at very high temperatures. Taq DNA polymerase is a highly thermostable recombinant DNA polymerase. It is named after *Thermus aquaticus*, the heat-tolerant bacterium from which it is isolated. *T. aquaticus* lives in hot springs and hydrothermal vents. Its DNA polymerase is very heat-stable and is most active at around 70 °C (a temperature at which a human or *E. coli* DNA polymerase would be nonfunctional). It amplifies DNA up to 5 kb and is stable even at 95 °C (van Pelt-Verkuil et al., 2008). Due to its crucial role in synthesizing and amplifying new strands of DNA at elevated temperatures, *Taq* DNA Polymerase is essential to PCR.

## 3  Primers

Like other DNA polymerases, *Taq* polymerase can only make DNA if it's given a primer. Primer refers to a small set of nucleotides of single-stranded DNA, typically 18–24 base pairs in length. These oligonucleotide sequences that will serve as primers for replication are chemically synthesized and complementarity with both ends of the target DNA, which is to be amplified. It is used in PCR to target a locus to allow for amplification for further analysis. It is crucial to have at least one pair of oligonucleotides for selective amplification of nucleotide sequences from a DNA sample by PCR. One of the primers is designed to recognize complementarily a sequence located upstream of the 5′–3′ strand of DNA, while the other to recognize a sequence located upstream complementary strand 3′–5′ of the same fragment of DNA. Two primers are used in each PCR reaction, and they are designed to flank the target region (a region that should be copied). They are given sequences that will make them bind to opposite strands of the template DNA, just at the edges of the region to be copied. The primers bind to the template by complementary base pairing. When the primers are bound to the template, they can be extended by the polymerase, and the region between them will be copied.

Calculation of melting temperature:

$$T_m \approx 4(G - C) + 2(A - T).$$

The following points should be kept in mind while designing a primer:

(a) The length of the primer should be in the range of 15–30 bases.
(b) G-C content should not exceed 40–60%.
(c) The 3′ end of primers should contain a G or C, and the 3′ ends of the primer set should not be complementary to each other.
(d) Optimal melting temperatures (Tm) for primers range between 50 and 65 °C; although the range can be broader or narrower, the final Tm for both primers should not differ by more than 5 °C.

(e) Dinucleotide repeats (e.g., GCGCGCGCGC or ATATATATAT) or single base runs (e.g., AAAAA or CCCCC) should be avoided (Lorenz, 2012).

1. **Nucleotides (dNTPs or deoxynucleotide triphosphates)**

These are single units of the bases A, T, G, and C, which are essentially "building blocks" for new DNA strands.

# 4 Buffer System

Magnesium and potassium provide optimum conditions for DNA denaturation and renaturation. It is also vital for fidelity, polymerase activity, and stability.

# 5 The Reaction Conditions

The conditions of any PCR-based reaction mainly include the total number of cycles and the temperature and duration of each step in those cycles. It is imperative to know how many cycles one must run. It depends on the amount of starting DNA and the number of copies of the PCR product required. A typical PCR reaction contains 25–35 cycles. This results in the generation of approximately 34 million to 34 billion copies of the template DNA. It is called a standard PCR run. Cycle numbers can be varied depending on the lower availability of starting DNA. However, reactions over 45 cycles are quite rare. Also, increasing the number of cycles for larger amounts of starting material is counterproductive because the presence of very high concentrations of the PCR product is itself inhibitory (Lorenz, 2012). Once the number of cycles is finalized, it is required to choose the temperature and time of each step in the cycles. The PCR consists of three defined sets of times and temperatures termed steps: (1) denaturation, (2) annealing, and (3) extension. Each of these steps is repeated 30–40 times, termed cycles (Fig. 2).

The first step is the **DNA denaturation step** that makes all of the DNA in the reaction single-stranded in the first cycle by heating the reaction at around 95 °C for 5 min initially and then for 30 s in each subsequent cycle. The second step is the **primer annealing step**, during which the PCR primers bind complementarily to the target. The annealing temperature is determined by the melting temperature (Tm) of the two PCR primers. It usually lies in the range of 55–62 °C. The exact temperature is critical for each PCR system and has to be defined and optimized. Optimization of annealing temperature is crucial; otherwise, non-precise reactions may give rise to other DNA products in addition to the specific target or may not produce any amplified products at all. One helpful technique for optimization is touchdown PCR, where a programmable cycler is used to decrease the annealing temperature

**Fig. 2** Different stages in
a single PCR cycle

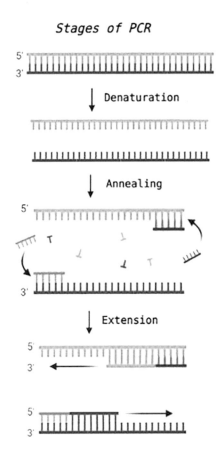

until the optimum is derived incrementally. The usual duration for annealing in each cycle is 30 s. Finally, the last step in a PCR cycle is the **polymerase extension step**, during which the DNA polymerase produces a complimentary copy of the target DNA strand (Fig. 2). The average temperature of this step is 72 °C. In addition to these steps, it is often recommended to add a single denaturation step of 3–5 min at 95 °C at the beginning of the reaction and a final extension step of 3 min at 72 °C. It is referred to as the final extension step.

As the reaction progresses through successive denaturation, annealing, and extension cycles, a single segment of double-stranded DNA template is amplified into two separate pieces of double-stranded DNA. These two pieces are then available for amplification in the next cycle. As the cycles are repeated, more and more copies are generated, and the number of copies of the template increases exponentially (van Pelt-Verkuil et al., 2008).

## 6   Basic Principle

The basic principle of PCR is similar to that of replication in vivo systems. Using the enzyme DNA polymerase, PCR aims to synthesize DNA from deoxynucleotides on a single-stranded DNA template. The polymerase chain reaction is carried out in a mixture comprising the template DNA, the forward and reverse primers, Taq polymerase, and the four deoxyribonucleoside triphosphates (dNTPs) in excess in a buffer mix. Thus, if a synthetic oligonucleotide is annealed to a single-stranded template containing a region complementary to the oligonucleotide, DNA polymerase can use it as a primer to elongate its 3′ end to generate an extended region of double-stranded DNA. The mixture reaction tubes are subjected to repetitive temperature cycles in automatic, programmed steps in a thermocycler (Kainz, 2000).

At first, the reaction mix is heated to unwound the double-stranded DNA template into single strands, referred to as denaturation. The temperature of the reaction mix is lowered to allow the binding of primers to the DNA templates. Thereafter, the elongation temperature (72 °C) favors the synthesis of new DNA strand using these primers with the help of DNA polymerase. This whole procedure of denaturation, primer binding, and synthesis completes the first cycle of PCR, and we get from each double-stranded DNA molecule one new and one old DNA strand. PCR then continues with further cycles that repeat the steps mentioned above. The newly produced DNA segments act as templates in later cycles, which allow the DNA target to be exponentially amplified millions of times (Kadri, 2019).

## 7   Short Protocol

Prerequisites for setting up a PCR reaction:

(a) Quantify the DNA of your interest.
(b) Dilute your primers to a working concentration.
(c) Standardize the annealing temperature and number of cycles.
(d) Make sure to have a PCR buffer that contains $MgCl_2$; if not, then $MgCl_2$ can be added separately.
(e) Decide the reaction volume (20 μL–100 μL).
(f) Detection system for checking your amplicons.

The standard protocol for setting up the reaction is as follows:

(a) Take a PCR tube (200 μL) and label it.
(b) Add forward and reverse primer to it, working concentration to be 10 pM.
(c) Add 10X PCR buffer to a final concentration of 1X.
(d) To the above mixture, add dNTPs and $MgCl_2$ solution.
(e) In the end, add the DNA template and Taq polymerase.
(f) Spin the tube to settle down anything on the walls and remove air bubbles.

(g) Now switch on the thermocycler; set the number of cycles and annealing temperature.
(h) Enter the volume of reaction and start the program.
(i) Once the reaction is complete, the PCR product may be stored at 4 °C or undergo electrophoresis for further applications.

Precautions

(a) All components of PCR should be kept on ice, and experiments should all be carried out on the ice.
(b) The concentration of MgCl2 is very critical for the reaction. It should also be standardized.
(c) Avoid contamination of the tubes, and to rule out any contaminants, always use positive and negative controls.
(d) Always prepare the master mix to avoid pipetting errors.

# 8 Applications

PCR in genetics research: PCR is used extensively for gene mapping, DNA sequencing, and analyzing the expression of genes under various conditions. Further, it is used to compare the genome of different organisms as well as phylogenetic analysis of DNA from unknown sources.

PCR in medicine: PCR is beneficial for the detection of mutants and disease-causing genes in the parents. It also helps in the testing of genetic diseases and monitoring of genes during gene therapy.

PCR in forensic science: PCR is immensely beneficial for paternity testing and identification of criminals using their DNA. Moreover, it is used as a tool in genetic fingerprinting.

# References

Kadri, K. (2019). *Polymerase chain reaction (PCR): Principle and applications, synthetic biology—New interdisciplinary science, Madan L. Nagpal, Oana-Maria Boldura, Cornel Baltă and Shymaa Enany.* IntechOpen. https://doi.org/10.5772/intechopen.86491

Kainz, P. (2000). The PCR plateau phase- towards an understanding of its limitations. *Biochem Biophys Acta, 1494*, 23–27.

Lorenz, T. C. (2012). Polymerase chain reaction: Basic protocol plus troubleshooting and optimization strategies. *Journal of Visualized Experiments, 63*, e3998. https://doi.org/10.3791/3998

Shampo, M. A., & Kyle, R. A. (2002). Kary B. Mullis—Nobel laureate for procedure to replicate DNA. *Mayo Clinic Proceedings., 77*(7), 606. https://doi.org/10.4065/77.7.606

van Pelt-Verkuil, E., van Belkum, A., & Hay, J. P. (2008). *Principles and technical aspects of PCR amplification.* Springer.

Wilson, K., & Walker, J. (2010). *Principles and techniques of biochemistry and molecular biology* (7th ed.). Cambridge university press.

# Southern and Northern Blotting

**Abstract** Blotting techniques are among the most common molecular biology approaches to immobilize proteins or nucleic acid (DNA/RNA) separation using a membrane. Samples can be detected by specific ligands, antibodies, or nucleic acid probes, which bind with individual nucleic acid sequences or proteins. These techniques depend upon the resolving ability of the individual macromolecules with respect to their size. The most well-recognized application of blotting is the Southern blot and Western blot, in which nucleic acid/proteins are immobilized on PVDF or nitrocellulose membranes. Blotting technology has been utilized over the last 30 years to elucidate many fundamental biological processes and promises more excellent discoveries in molecular biology. This chapter discussed the principle and methodology of the different types of blotting techniques.

**Keywords** Blotting · Nitrocellulose membrane · Gel electrophoresis · Hybridization · Probe

## 1 Introduction

Blotting is a gel electrophoresis method for separating DNA, RNA, and proteins. Southern blotting is used to detect DNA molecules, and Northern blotting is used to detect RNA from the sample mixture post-separation on the gel. Both the techniques are almost similar, with minor variations. In both, DNA/RNA molecules from the agarose gel are transferred to the nylon or nitrocellulose membrane using buffer-soaked Whatman filter paper or absorbent paper (Sambrook et al., 1989). Then, a specific sequence of DNA or RNA can be detected by molecular hybridization with radioactive labeled DNA or RNA probe ($P^{32}$, $H^3$, $S^{35}$ or $I^{125}$) or using fluorescent and chemiluminescent reagents (Denhardt, 1966; Botchan et al., 1976), while proteins are detected by labeled antibodies (Southern, 2006). Since mid 1990s, Blotting has been regularly used to study genetic mapping of humans (Botstein et al., 1980; Kan & Dozy, 1978), rabbit genome sequencing (Jeffreys &

© The Author(s), under exclusive license to Springer Nature Switzerland AG 2022     165
A. Gautam, *DNA and RNA Isolation Techniques for Non-Experts*, Techniques in
Life Science and Biomedicine for the Non-Expert,
https://doi.org/10.1007/978-3-030-94230-4_21

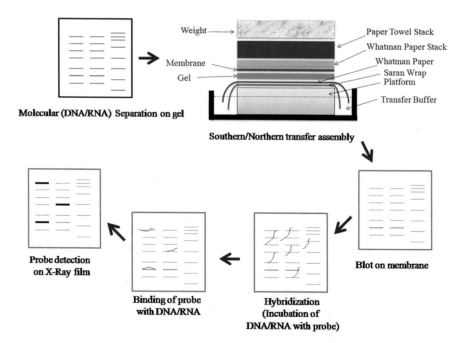

**Fig. 1** Steps involved in Southern/Northern blotting

Flavell, 1977), DNA fingerprinting (Jeffreys et al., 1985), and DNA/RNA structure (Brewer & Fangman, 1987).

There are several steps involved in Southern and Northern blotting (Fig. 1), including (1) purification of the DNA/RNA, (2) digestion with restriction enzyme (not required for RNA), (3) gel electrophoresis, (4) membrane transfer, (5) probe preparation and labeling, and (6) detection. For Southern and Northern blotting, two types of membrane (nitrocellulose/nylon) are required. The membranes are made of uncharged or positively charged nylon or nitrocellulose. Nowadays, fragile nitrocellulose membranes have been replaced by nylon membranes for specific applications. The main advantage of using nylon membranes is that they can be reprobed more than ten times compared to the nitrocellulose membranes, which can be reprobed not more than three times.

The Southern/Northern blot procedure to blot gel onto the nylon membrane (positively charged or uncharged) has been discussed below.

## 2   Materials Required

TAE (Tris acetic acid) or TBE (Tris borate) buffer
  Agarose
  Ethidium bromide (EtBr) or SYBR Green

2X and 20X SSC (0.3 M sodium citrate and 3.0 M NaCl)
6X loading buffer
Molecular markers
Prehybridization and hybridization solution (0.5% SDS, 6XSSC)
Denhardt's solution, or 10 mg/mL denatured, sheared salmon sperm DNA
Paraffin oil
Nylon membranes
RNase A
Distilled water (DW)
Buffer and restriction enzyme
RNA (radioactively labeled)
2,5-Diphenyloxazole (PPO) in toluene

# 3  General Protocol

## 3.1  Restriction Digestion and Gel Electrophoresis

1. Digest the DNA samples with restriction enzyme(s), perform agarose gel electrophoresis with molecular marker, and observe the gel's band stain with EtBr. There is no need to digest the RNA sample as it can be separated on the gel using a denaturing agent that confines the secondary RNA structures.
2. For the depurination, rinse the gel with distilled water (DW) and keep the gel in a glass dish, and submerge in 0.25 M HCl.
3. Keep on a shaker for 30 min at room temperature (RT).
4. After 30 min, rinse the gel with DW, and keep it in denaturation solution for 20 min on the shaker (to unwind the DNA to a single strand). Repeat this step twice for proper denaturation.
5. Rinse the gel with DW and add neutralization solution to reduce the pH of gel <9.0 (higher pH will not allow the DNA to bind with nitrocellulose). Keep on the shaker for 20 min and repeat this step twice.

## 3.2  Transfer

6. The transfer requires one piece of thick filter paper. The size of the paper should be more than gel and should be presoaked in 20× SSC buffer. Similarly, we need one piece of nitrocellulose membrane soaked in 2× SSC buffer before transfer. Assemble the components for transfer as illustrated in Fig. 1.
7. Keep soaked filter paper (in 20XSSC buffer) on the platform, place gel on it, and remove air bubbles from the area between gel and platform by using a glass pipet to roll over the surface of the gel.

8. Cover the edges of the gel with plastic wrap to prevent short-circuiting.
9. Place the soaked membrane (in 20× SSC) on the exposed surface of the gel, and remove the air bubbles with the help of a glass pipette.
10. Pour 20× SSC on the membrane, and keep 5–10 sheets of filter paper the same size as the membrane.
11. Keep the membrane-sized paper towel on the top of filter paper.
12. Place a glass plate over the paper towel, and place some weight on it to ensure good contact between the stack.
13. Leave it overnight.
14. Transfer time can be extended for >1% agarose gel and desired fragment size >20 kb.
15. Disassemble the blotting transfer setup.
16. Recover the membrane and mark with pencil faced toward gel to recognize back portion of the membrane.
17. Rinse the membrane in 2× SSC buffer, and dry with the help of Whatman 3 mm paper sheets.
18. To immobilize the DNA/RNA, place the membrane between two sheets, wrap with UV-transparent plastic sheets, and place DNA-side-down on a UV transilluminator or keep it in the oven for 2 h at 80 °C.
19. After this step, the membrane can be stored between filter paper sheets for several months at room temperature or in a desiccator at room temperature or 4 °C.

## 3.3   Prehybridization

20. The prehybridization (blocking) process is used to diminish nonspecific binding between probe and membrane.
21. A blocking agent (Denhardt's solution or Salmon sperm DNA) is used to prevent probe binding to the membrane.
22. Pour the prepared prehybridization solution into the hybridization tube/box, and keep it at 42 °C in a hybridization oven for 5 h.

## 3.4   Hybridization

23. In this step, the desired sequence is detected by DNA/RNA probe.
24. Incubate the probe mix in the water bath for 40 min at 37 °C.
25. Further incubation will be performed again with hybridization solution overnight at 49–52 °C.
26. Warm the 6X washing solution at 49 °C, and add to the hybridization tube after discarding the hybridization solution.

## 3.5   *Detection of Probe*

27. Use an X-ray film cassette and wrap the blot carefully. Keep the cassette in the darkroom at −80 °C for overnight to develop the film.

# 4   Application

Southern blotting techniques are regularly used in molecular biology studies, such as identifying restriction fragments containing DNA sequence or a gene of interest, constructing genomic maps, etc. The northern blotting technique is used to observe a particular gene's expression pattern in tissues/organs during environmental stress levels, developmental stages, and pathogen infection. These techniques have been used to determine the overexpression of oncogenes and downregulation of tumor-suppressor genes.

# References

Botchan, M., Topp, W., & Sambrook, J. (1976). The arrangement of simian virus 40 sequences in the DNA of transformed cells. *Cell, 9*, 269–287.

Botstein, D., White, R. L., Skolnick, M., & Davis, R. W. (1980). Construction of a genetic linkage map in man using restriction fragment length polymorphisms. *American Journal of Human Genetics, 32*, 314–331.

Brewer, B. J., & Fangman, W. L. (1987). The localization of replication origins on ARS plasmids in *S. cerevisiae. Cell, 51*, 463–471.

Denhardt, D. T. (1966). A membrane-filter technique for the detection of complementary DNA. *Biochemical and Biophysical Research Communications, 23*, 641–646.

Jeffreys, A. J., & Flavell, R. A. (1977). The rabbit β-globin gene contains a large large insert in the coding sequence. *Cell, 12*, 1097–1108.

Jeffreys, A. J., Wilson, V., & Thein, S. L. (1985). Individual-specific 'fingerprints' of human DNA. *Nature, 316*, 76–79.

Kan, Y. W., & Dozy, A. M. (1978). Polymorphism of DNA sequence adjacent to human beta-globin structural gene: Relationship to sickle mutation. *Proceedings of the National Academy of Sciences of the USA, 75*, 5631–5635.

Sambrook, J., Fritsch, E. F., & Maniatis, T. (1989). *Cold Spring Harbor*. Cold Spring Harbor Laboratory Press.

Southern, E. (2006). Southern blotting. *Nature Protocols, 1*(2), 518–525.

# Genome Mapping

**Abstract** The isolated DNA and its sequencing are one of the essential steps in genome mapping. Genome mapping is a method to mark the physical locations of different genes on a chromosome relative to each other. During genome mapping, different available or known molecular markers on genes like RAPD, regulatory factors, DNA polymorphism, etc. are used as landmarks. Based on the basis by which the distance between two molecular markers is determined and then plotted on a gene in a chromosome, genome mapping can be of two types: genetic mapping and physical mapping. This chapter discusses the difference between these two types as well as different subtypes of genome mapping and their application in genetics and molecular biology.

**Keywords** Genes · Chromosomes · Mapping · Sequencing · Molecular marker · DNA

## 1 Introduction

The genome mapping methods determine the exact location of genes on a chromosome and their relative distance from other genes or specific regions of the chromosome (like centromere and telomere). The different genome mapping methods can be categorized into genetic (or linkage) mapping and physical (or non-genetic) mapping. Genome mapping helps identify genes responsible for causing heritable diseases (like cystic fibrosis, sickle-cell anemia) and other genetic diseases (like cancer). Moreover, genome mapping is widely used to produce genetically modified organisms by selecting and propagating the desirable traits in animals and plants, e.g., disease resistance, meat production, high-yielding variety seeds, etc. Genetic mapping has practical implications in crime investigations, paternity tests, gene therapy, organ donor and recipient mismatch, etc. The different genome mapping

A. Gautam, *DNA and RNA Isolation Techniques for Non-Experts*, Techniques in Life Science and Biomedicine for the Non-Expert, https://doi.org/10.1007/978-3-030-94230-4_22

methods and their applications have been discussed in detail in this chapter. Before that, the clarification between DNA sequencing and genome mapping needs to be addressed.

## 2 Difference Between Genome Mapping and DNA Sequencing

DNA sequencing involves the readout of A, T, G, C base pairs in a particular DNA fragment, rather than the whole genome or chromosome. In one sense, DNA sequencing is a part of genome sequencing, as the latter technique involves the physical reconstruction of the whole genome from the pieces of information gathered by sequencing. For example, the information of genes, unique sequences, restriction sites, repetitive sequences, etc. through DNA sequencing is utilized to provide a detailed picture of the genome organization in genome mapping. Such arrangement helps in plotting the gene order and their relative distance precisely in a genome.

## 3 Types of Genome Mapping

Based on the basic principle applied to determine the arrangement of DNA markers and genes in a chromosome, genome mapping methods are of two types: genetic mapping and physical mapping. Genetic mapping relies on the recombination frequencies, whereas physical mapping depends on the number of nucleotide pairs between loci to mark the relative distance between positions. Due to this difference, genetic mapping is also known as indirect or linkage mapping, and physical mapping is as direct or non-genetic mapping. The principal genetic mapping method utilizes either genes, restriction fragment length polymorphism (RFLP), or molecular markers. Physical mapping can be done by fluorescence in situ hybridization (FISH), karyotyping, radiation hybridization, STS content mapping, and restriction mapping.

## 4 Genetic Mapping

For the construction of a genetic map, genes and molecular markers on a chromosome (like RFLP, RAPD, AFLP) act as landmarks. The recombination frequency (obtained during crossovers or breeding experiments) of these landmarks helps speculate their distance (in centimorgan) on the genome. For example, very closely located genes (linked genes) are always inherited together, whereas distantly located

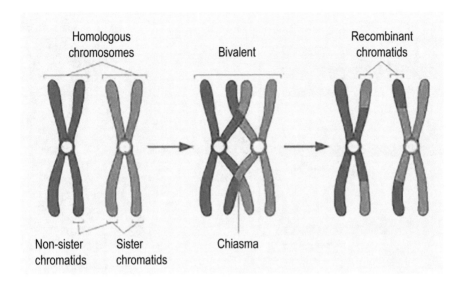

Homologous chromosomes    Bivalent    Recombinant chromatids

Non-sister chromatids    Sister chromatids    Chiasma

**Fig. 1** Pairing of two homologous chromosomes occurs in meiosis. After that, the crossing-over between two non-sister chromatids takes place through the formation of a chiasma. This crossing-over results in the formation of recombinant chromatids

genes (non-linked genes) generally get separated after the crossing-over. However, due to the absence of molecular markers, unavailability of recombination frequency data, and the presence of junk DNA in the genome, the genetic mapping is constructed with limited landmarks and therefore is considered a low-density mapping.

*Mapping by known location of genes*: During meiosis, non-sister chromatids break off and recombine with the same strand of the other homolog (Fig. 1) (Saraswathy and Ramalingam, 2011). This recombination is observed in the phenotype of the progeny after the breeding experiments (or pedigree data for humans), and after that, the percentage of recombination frequency (i.e., the ratio of the recombinant progenies) and LOD (logarithm of the odds) score can be calculated easily. The recombination frequency is directly proportional to the distance between two genes of a chromosome and their linkage. A positive LOD score represents the linked genes, and a negative LOD score is obtained for non-linked genes. Such type of mapping is also known as linkage mapping and does not represent the actual physical distance between two genes.

*Mapping by known location of molecular markers*: The presence of different DNA sequences in individuals of a species due to single/multi-base pair change and repetitions of sequences is known as DNA polymorphism. DNA polymorphism provides genomic variability as well as serves as the molecular marker for its own location in a chromosome. Some widely studied DNA polymorphisms are single nucleotide polymorphisms (SNPs), transposable elements (e.g., maize genome), minisatellites, etc. These can be detected through molecular techniques like PCR, Southern blots, and DNA microarray chips (Mohan et al., 1997). Thereafter, their

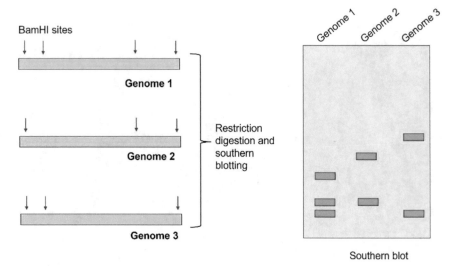

**Fig. 2** Three different genomes with different restriction sites for the BamHI enzyme will yield different restriction fragments after the restriction digestion. The Southern blot detects this restriction fragment length polymorphism (RFLP) after DNA gel electrophoresis. Based on the molecular weight of detected fragments, one can construct a genetic map for this gene

inheritance pattern is examined in the following progenies by either breeding experiments or pedigree analysis. These observations are used to construct a genetic map with the help of computer programs. One of the commonly used polymorphisms for constructing a genetic map is RFLP (restriction fragment length polymorphism), which is based on the presence or absence of the restriction sites (Saraswathy and Ramalingam, 2011). The brief protocol has been depicted in Fig. 2.

## 5 Physical Mapping

Physical mapping provides us a high-density genome map as they directly pinpoint the DNA sequences in a chromosome. In the case of physical mapping, the unit of measurement is base pairs instead of a centimorgan (1 centimorgan ≈ 1 million base pairs). Harnessing the availability of different markers on a chromosome such as unique DNA sequences, EST (expressed sequence tags), and STS (sequence tagged site) markers, one can construct a physical map with the help of commonly used molecular techniques like Southern hybridization, FISH, PCR, restriction analysis, etc. The following are the major types of physical mapping:

*Karyotyping*: Karyotyping is a cytogenetic method to stain an individual's chromosomes using specific dyes and then observe their number, size, and banding patterns. The two most commonly used dyes are Giemsa stain and fluorescent quinacrine. The former one stains the chromosome either in dark regions

**Fig. 3** Banding pattern observed in chromosomes after G-staining

(heterochromatic) or light regions (euchromatic) to give a banding pattern. This is known as G-banding (Fig. 3). On the other hand, Q-banding is produced similarly by using quinacrine dye. This fluorescent dye binds to the AT-rich DNA region and forms an AT-quinacrine complex. Since this complex fluoresces in the presence of suitable electromagnetic radiations, we can see a banding pattern of pale light (by GC-rich regions) and dark yellow light (due to AT regions), very similar to G-banding. The abnormalities in the banding patterns, which lead to visible chromosomal diseases or genetic disorders, are the source for the physical mapping of affected genes (Altshuler et al., 2008).

*Restriction enzyme mapping*: Several DNA sequences are recognized by specific endonucleases and get cleaved. These sequences are known as restriction sites, and the enzymes involved in this restriction digestion are called restriction enzymes. Therefore, the basic principle of restriction enzyme mapping involves using known restriction enzymes at unmapped DNA stretch and observing the cleavage site after the enzymatic action (Fig. 4). Though this method is suitable for smaller genomes (as of bacteria and viruses), the production and analysis of millions of cleaved DNA fragments in larger genomes (e.g., human genome) makes this method tedious and unfitting. However, with the advent of automated methods and optical mapping, restriction enzyme mapping can also be done for larger genomes (Altshuler et al., 2008).

*Fluorescent* in situ *hybridization*: Fluorescence in situ hybridization (FISH) is a cytogenetic technique to locate the specific sequence or gene on the metaphase chromosome through the fluorescent-labeled probes. In this technique, metaphase chromosomes are allowed to expose their sites by denaturation of proteins and then hybridized with fluorescent-tagged cDNA or EST probes. Finally, the complementary binding of these probes and hence the location of specific sequences can be observed through a fluorescent microscope (Fig. 5). As this technique provides a direct association between the probe sequence and chromosome, it is widely used for assembling genes directly on the chromosome.

*Radiation hybrid mapping*: Radiation hybrid mapping is the most efficient method of genome mapping. It can generate a high-resolution map (even of the

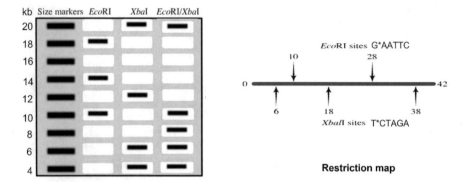

**Fig. 4** During the restriction mapping, an unknown stretch of DNA is digested with a combination of several restriction enzymes (e.g., Eco RI and Xba I). After proper digestion, the digested fragments are electrophoresed on the agarose gel, and their length is determined based on their molecular weight. A restriction map is constructed by assembling all the fragments at the proper place to construct a full-length initial DNA stretch

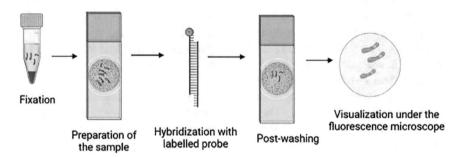

**Fig. 5** For FISH, metaphase chromosomes are fixed and then denatured by formaldehyde to make them single-stranded at the glass slide. Fluorescence-labeled probes are allowed to hybridize onto this glass slide. Several rounds of washings remove nonspecific bindings with buffer solutions. The slide is visualized under a fluorescence microscope post-washing to locate the position of the hybridized probe in the whole chromosome

whole human genome) using polymorphic and non-polymorphic markers. It involves introducing breaks in the targeted chromosome (which needs to be mapped) by irradiation of cells. These cells with fragmented chromosomes are implanted in rodent cells and allowed to grow as a hybrid cell using appropriate selection medium (e.g., HAT medium). Then, different markers (STS, EST, or non-polymorphic markers) can be located in these hybrid cells using PCR amplification. The linkage between the two markers is established by analyzing the breakage percentage. The distantly located markers will be detected in distinct DNA fragments due to easy breakage by irradiation. In contrast, the closely situated markers will be detected in

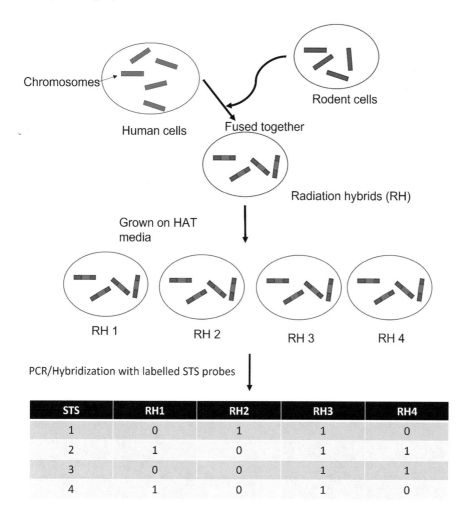

**Fig. 6** During radiation hybrid mapping, chromosomes to be mapped are irradiated to generate different length fragments. These fragmented chromosomes are fused with rodent chromosomes and propagated as radiation hybrids (RH) in the presence of HAT media. With the help of PCR or Southern hybridization, each RH cell line is examined for the presence (scored as 1) or absence (scored as 0) of an STS marker. The relative location of the STS markers and their overlapping clones are utilized for the physical mapping

the same fragment due to retention in the same fragment (Fig. 6). The frequency of the breakage of the markers into different fragments is utilized to assess the chromosomal distance between them (Lawrence et al., 1991).

# References

Altshuler, D., Daly, M. J., & Lander, E. S. (2008). *Genetic mapping in human disease. science, 322*(5903), 881–888.

Lawrence, S., Morton, N. E., & Cox, D. R. (1991). Radiation hybrid mapping. *Proceedings of the National Academy of Sciences, 88*(17), 7477–7480.

Mohan, M., Nair, S., Bhagwat, A., Krishna, T. G., Yano, M., Bhatia, C. R., & Sasaki, T. (1997). Genome mapping, molecular markers and marker-assisted selection in crop plants. *Molecular Breeding, 3*(2), 87–103.

Saraswathy, N., & Ramalingam, P. (2011). *Concepts and techniques in genomics and proteomics.* Elsevier.

# Applications of DNA Sequencing Technologies for Current Research

**Abstract** In genomic analysis, next-generation sequencing (NGS) technology continues to bring cost-effectiveness, unparalleled sequencing speed, with high promise, and accuracy. NGS technology is the next stage in the evolution of DNA sequencing, allowing for the creation of hundreds to millions of DNA sequences in a short period. The relatively rapid adoption and effectiveness of NGS in research transformed genomes and medical diagnosis. The current method of medical diagnosis has given way to the personalized medicine paradigm, which allows for more accurate diagnosis of human ailments as well as the identification of molecular targets for individualized treatment. Furthermore, NGS approaches provide genome-wide profiling and characterization of mRNAs, short RNAs, transcription factor regions, chromatin structure, whole-genome sequencing, target sequencing, and chromatin sequencing. In addition, NGS has been widely employed to speed up biological and biomedical research in several ways. In this chapter, we address the technological foundations of numerous commercially available NGS systems in this chapter, as well as the current state, technique, and comparison of first-, second-, third-, and fourth-generation sequencing technologies.

**Keywords** Next-generation sequencing · De novo sequencing · Shotgun sequencing · Illumina sequencing · Ion Torrent semiconductor sequencing

## 1 Introduction

DNA sequencing techniques have transformed biomedical research. Sequencing techniques with enhanced sensitivity and throughput are much required for the molecular and genomics study (Diaz-Sanchez et al., 2013; Buermans & den Dunnen, 2014; Bansal et al., 2018). Four building blocks (A, T, G, C) known as the nitrogenous base are involved in DNA synthesis, and their orderly arrangement to make a protein is known as a gene. Many severe and new challenges related to public health and safety have emerged because of the advancement of DNA sequencing tools.

During DNA sequencing, each nucleotide is recorded as signals accumulated by the machine and studied on the computer. Universities, companies, and institutions are constantly working to improve DNA sequencing applications, namely, whole-genome sequencing, custom target sequencing, and whole-exome sequencing, which are cost-effective and efficient diagnostic methods (Hogan, 2006; Barba et al., 2014; Nimse et al., 2014; Reuter et al., 2015).

DNA sequencing has revolutionized the biological sciences with its wide range of applications. Researchers can gather data about genetic variation and gene expression patterns on a massive scale because of the rapid advancement of DNA sequencing technologies (Buermans & den Dunnen, 2014; Bansal et al., 2018). These technologies are anticipated to make accurate sequencing of complete human genomes a standard tool for researchers and clinicians in the coming years. Certain other areas dependent on DNA sequencing are drug trials and pharmacogenomics, newborn and pediatric disease, rare tumor type, regulatory variation and eQTLs, clan genomics (Roden et al., 2011), large cohorts with extensive phenotyping, and family disease pedigrees. However, researchers seeking to take full advantage of these rapidly developing technologies will need to be attentive to the challenges ahead, particularly in terms of the communications and expertise required for effective management of a large amount of data obtained using the new sequencing platforms (McCarty et al., 2005; Zhao et al., 2017).

Large-scale DNA sequencing requires robotized techniques dependent on fluorescence labeling of DNA and suitable identification frameworks (Head et al., 2014). For the most part, a fluorescent tag can be utilized either directly or in an indirect way. Direct fluorescent marks, as utilized in mechanized sequencing, are fluorophores. These are atoms that transmit different fluorescent light when exposed to UV light of a particular frequency. A laser is utilized to invigorate the fluorescent color rather than utilizing X-ray film to examine the arrangement. The base-explicit fluorescent marks are joined to suitable dideoxynucleotide triphosphates (ddNTP). Each ddNTP is marked differently (Zhou & Li, 2015; Deharvengt & Tsongalis, 2018). The laser initiates a fluorescent sign subject to the particular tag addressing one of the four nucleotides. The fluorescent discharges are gathered on a charge-coupled gadget that can verify the wavelength.

## 2   History

In 1970 at Cornell University, Ray Wu developed the first DNA sequencing method based on a position-specific primer-extension approach (Xue et al., 2016). The cohesive ends of the lambda phage DNA were sequenced by DNA polymerase and particular nucleotide labels, both typical for current sequencing methods (Heather & Chain, 2016). In 1973, Wu, R Padmanabhan, and colleagues stated that synthetic location-specific primers approach could identify any DNA sequence (Hughes & Ellington, 2017). Frederick Sanger used this primer-extension method for rapid DNA sequencing at the MRC Centre in Cambridge, UK, and in 1977 also published a report on "DNA sequencing using chain-terminating inhibitors" (Barba et al., 2014). At Harvard, two researchers Maxam and Gilbert devised sequencing techniques for DNA sequence determination by chemical degradation (García-Sancho, 2010).

The first-generation sequencing strategy was based on a chain-termination approach proposed by F. Sanger and colleagues, which revolutionized the world of biological sciences (Sanger et al., 1977). Interestingly, in 1977, Allan Maxam and Walter Gilbert explained the chemical method of DNA sequencing (in bacteriophage X174) (Heather & Chain, 2016). In 1986, Lorey and Smith proposed the first semiautomated DNA sequencing method. Later on, in 1987, the Applied Biosystems developed a fully automated machine-controlled DNA sequencing method. After the development of fully automated machines, the era of the 2000s becomes an unprecedented era for sequencing platforms. Afterward, in 2003, the Human Genome Project was completed using this method (Liu et al., 2012; Heather & Chain, 2016). Solexa/Illumina recognized a quick, accurate, reliable, and highly competent next-generation sequencing platform in 2005 (Shokralla et al., 2012).

## 3   Basic Principle

The method of DNA sequencing determines the exact sequencing order of nucleotides bases in the extracted DNA portion (Heather & Chain, 2016) (Fig. 1). Two distinct procedures for DNA sequencing have been proposed: the chain-termination technique and the chemical degradation technique known as the first-generation sequencing approach (Mitchelson, 2005). Both methods had equal acceptance, to begin with, but, for various reasons, the former one is more often in practice nowadays. For this achievement, Frederick Sanger was awarded the 1980 Nobel Prize in chemistry (Shetty et al., 2019). Therefore, we can categorize basic methods of DNA sequencing into two following heads:

1. Maxam and Gilbert chemical degradation method
2. Chain-termination or dideoxy method Fredrick Sanger

## 4   Maxam-Gilbert Sequencing

Maxam and Gilbert described a sequencing method based on chemical degradation at precise locations of the DNA molecule (Gupta & Gupta, 2014). The end-labeled DNA fragments are subjected to random cuts at adenine, cytosine, guanine, or thymine positions using specific chemical agents. Polyacrylamide gel electrophoresis (PAGE) separates the products of these four reactions (Maxam & Gilbert, 1986). The sequence can be straightforwardly read from four parallel lanes in the sequencing gel per the Sanger method. As in the Sanger method, additional attentiveness in

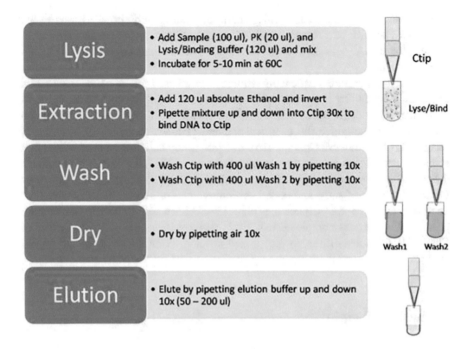

**Fig. 1** Basic protocol for DNA extraction

Maxam and Gilbert method incorporates purification and separation of DNA fragments and takes higher analysis time. Therefore, this technology is not satisfactory for high-throughput large-scale investigation.

## 5 Chain-Termination or Dideoxy Method/Sanger Sequencing Principle

Sanger sequencing involves producing multiple copies of a target DNA region (Sanger et al., 1977). Its ingredients are like those needed for DNA replication in an organism or for polymerase chain reaction (PCR), which copies DNA in vitro (Clark et al., 2019). The reaction mixture consists of a DNA polymerase enzyme, a primer tagged with a mildly radioactive molecule, or a light-emitting chemical. A short sequence of single-stranded DNA binds to the template DNA, and it also acts as a "starter" for the polymerase (Berg et al., 2002). However, a Sanger sequencing reaction also contains a unique ingredient: dideoxy or chain-terminating versions of all four nucleotides (ddATP, ddTTP, ddCTP, ddGTP), each labeled with an assorted color of dye. Dideoxynucleotides are like normal deoxynucleotides, with one of the key differences is that they lack a hydroxyl group on the 3′. The 3′ hydroxyl group works as a "hook" that permits another nucleotide to be added to a current arrangement in a normal nucleotide.

Once a dideoxynucleotide is added to the chain elongation, no hydroxyl exists, and no further nucleotides can be added. The chain closes with the dideoxynucleotide, which is set apart with a specific shade of color. The restricted measure of a dideoxynucleotide is added to every response, permitting various responses to continue for different periods until, by some coincidence, DNA polymerase embeds a dideoxynucleotide, ending the response. Thus, the outcome is a bunch of new chains, all of the varying lengths (Hegedüs et al., 2018). To examine the recently produced chain, the four responses are run simultaneously on a polyacrylamide sequencing gel. The groups of atoms created within sight of ddATP are stacked into one path of the gel, and the other three families, produced with ddCTP, ddGTP, and ddTTP, are stacked into three neighboring paths. After electrophoresis, the DNA arrangement can be perused straightforwardly from the places of the groups in the gel.

Developing the proficiency of DNA sequencing requires the improvement of frameworks, which decline the requirement for manual activities by incorporating layout planning, sequencing responses, item division, and acknowledgment (Heather & Chain, 2016). A semi-mechanized framework, in which PCR-intensified biotinylated genomic or plasmid DNA is immobilized on streptavidin-covered attractive dots, has been created (Nimse et al., 2014).

## 6 Large-Scale Sequencing and De Novo Sequencing

Before the introduction of next-generation sequencing (NGS) technologies, nucleic acid sequences were determined by the clone-by-clone method due to high throughput and read length limitation (Reuter et al., 2015). DNA is processed for fragmentation by restriction enzymes in the clone-by-clone method, and each fragment is combined into a vector (van Holde & Zlatanova, 2018). This vector insert is transferred into a host cell for in vivo cloning. Each vector insert is extracted from the host cell and then amplified using PCR. This in vitro product is used for sequencing, and thus all fragments are sequenced and then assembled as a genome. The limitation of this clone-by-clone method has paved the way for developing robust technology for nucleic acid sequencing (Alberts et al., 2002).

## 7 Shotgun Sequencing

The process of whole-genome shotgun sequencing is much faster than clone-by-clone sequencing (Leamon & Rothberg, 2009). Moreover, if there is an existing reference sequence, this sequencing method is particularly efficient. It is much easier to collect a short DNA sequence by aligning it to an existing reference genome.

For this, the starting DNA is broken up randomly into lots of smaller pieces and arranged in a shotgun fashion, with each of those pieces sequenced independently. The resulting "sequence reads" generated from the different pieces are then analyzed by a computer program. This computer process is used repeatedly, ultimately yielding the complete sequence of the starting piece of DNA. The initial random fragmenting and reading of the DNA is termed "whole-genome shotgun sequencing" (Choudhuri, 2014; Reuter et al., 2015; Heather & Chain, 2016).

## 8    Clone Contig Approach

The clone contig sequencing method is a conventional method for sequencing a large genome (Clark & Pazdernik, 2013). The Human Genome Project is the best example of this sequencing method. To date, no platform can sequence a total length of DNA as a single read, so sequencing is carried out by fragmentation of nucleic acid (Amarasinghe et al., 2020). In this approach, DNA is subjected to fragmentation by two or more restriction enzymes to produce large fragments (a few KB to MB in size) with overlapping sequences. Each DNA fragment is subjected for cloning in large insert vectors such as BAC, YAC, and COSMIDS (Hogan, 2006; Clark & Pazdernik, 2013). These clone contigs are assembled to get a complete genome sequence. Every single contig is large enough to can't be sequenced on any platform in a single read, so again every single read is fragmented with two or more restriction enzymes to get overlapping fragments (Head et al., 2014). These fragments are now sequenced using the shotgun sequencing method, and the de novo approach assembles generated sequences. In this way, each clone contig is sequenced, and overlapping sequences now assemble these sequences.

## 9    Amplicon Sequencing

Amplicon sequencing is a tool for the study/analysis of the particular region of interest in nucleic acid. It is a precise method for analyzing genetic variation in specific genomic regions. The use of ultra-deep sequencing of PCR products facilitates variant identification and characterization (amplicons) (Anderson & Schrijver, 2010). NGS is used after oligonucleotide probes have been used to target and capture regions of interest (Gaudin & Desnues, 2018). Amplicon sequencing can aid in the detection of rare somatic mutations in complex samples (for instance, tumors with germline DNA). Another popular application is sequencing the bacterial 16S rRNA gene across different species, a widely used approach for phylogeny and taxonomy investigations, particularly in diverse metagenomics samples.

## 10 High-Throughput Methods

In the mid-1990s to late 1990s, several new technologies for DNA sequencing were invented, and by the year 2000, they had been integrated into commercial DNA sequences (Kircher & Kelso, 2010). These were collectively dubbed "NGS" or "second-generation" sequencing approaches to distinguish them from earlier methods, such as Sanger sequencing. NGS technology is often characterized as highly scalable, allowing the entire genome to be sequenced at once, in contrast to first-generation sequencing. This is usually performed by fragmenting the genome into small parts, randomly sampling for a fragment, and sequencing it using one of the technologies outlined below (Anderson & Schrijver, 2010). A whole genome may be sequenced because many sections are sequenced simultaneously in an automated process (thus the name "massively parallel" sequencing).

High-throughput sequencing technologies have brought a revolution in molecular analysis of the biotic world. The technology provides enormous sequence data and huge genetic information, which is buried in specific sequences (Pasipoularides, 2017). In terms of microbial population study, the high-throughput sequencing and bioinformatics tools provide a better understanding of phylogenetic affiliation and the functional role of community members. High-throughput sequencing is referred to as massively parallel sequencing, which provides immense sequence data in minimum time. Various platforms are available for high-throughput sequencing, and some of them are listed in Table 1 (Fig. 2).

NGS technology has greatly aided researchers in their quest for health insights, anthropologists in their inquiry into human origins, and igniting the "personalized medicine" program (Suwinski et al., 2019). However, it has also increased the margin for error. There are a variety of software programs available for computing NGS data, each with its algorithm. Even within one software program, the parameters can alter the conclusion of the analysis. Furthermore, the huge amounts of data generated by DNA sequencing necessitated the development of novel sequence analysis methods and tools. To address these problems, many efforts are made to build standards in the NGS area, the majority of which have been small-scale projects emanating from individual laboratories (Jennings et al., 2017). The BioCompute standard is the result of a huge, well-organized, FDA-funded effort.

## 11 Long-Read Sequencing Methods: Nanopore Sequencing

It is a third-generation approach to nucleic acid sequencing (Xiao & Zhou, 2020). Utilizing nanopore sequencing, a solitary particle of DNA or RNA can be sequenced without the requirement for PCR enhancement. Nanopore sequencing is a novel, adaptable innovation that empowers immediate, continuous examination of long DNA or RNA pieces. It works by observing changes in an electrical flow as nucleic acids pass through a protein nanopore. The subsequent signal is decoded to give the

**Table 1** Various platforms for high-throughput sequencing

| Platform | Reads/unit (single reads) | Max read length (bp) | Read type | Highlights |
|---|---|---|---|---|
| Oxford | 500 bp–2.3 Mb | Up to 4,000,000 | Single end | Longest read length of any NGS platform |
| Illumina | 4,000,000– 2,000,000,000 | 150–600 | Single end and pair end | Highest output of any benchtop sequencer (up to 1.5 TB, 5 billion pair end)<br>This platform is ideally suited for exome, transcriptomics, whole-genome, and targeted sequencing<br>Well suited for de novo and resequencing of small and large genomes |
| Ion Torrent | 400,000– 60,000,000 | 200–400 | Single end | Fast turnaround time, optimal for small genomes or targeted sequencing<br>Ion proton has a greater number of reads but shorter read lengths. Compared to Illumina (MiSeq), has a fewer number of reads and a shorter read length |
| PacBio | 22,000–47,000 | 15,000– 20,000 | Single end | Long read length sequencing platform |
| Roche 454 | 70,000–700,000 | 400–700 | Single end | Long read lengths make it ideal for sequencing of small genomes |
| SOLiD | 81,500,000– 266,666,667 | 100 | Single end and pair end | High throughput, good for resequencing. Short reads make it nonideal for de novo assembly |

particular DNA or RNA sequence arrangement. Nanopore sequencing can generally offer minimal expense genotyping, high portability for testing, and quick preparation of tests with the capacity to show constant changes.

## 12   Short-Read Sequencing Methods: 454 Pyrosequencing

Pyrosequencing is a technique for DNA sequencing that recognizes light produced during the expansion of nucleotides through the blend of a reciprocal strand of DNA (Barba et al., 2020). Pyrosequencing can be valuable for recognizing DNA and RNA from pretty much every source and has applications in microorganism's hostility screening, drug advancement, epigenetics, and numerous different fields of science (Barba et al., 2020). Pyrosequencing is used to expose the exact genetic code of a segment of DNA. It will also be used to detect single nucleotide polymorphisms, insertion-deletion, or other sequence variations, along with quantifying DNA methylation and allele frequency (Delaney et al., 2015).

**First Generation: Sanger Sequencing**

**Second Generation: Amplified Molecule Sequencing**

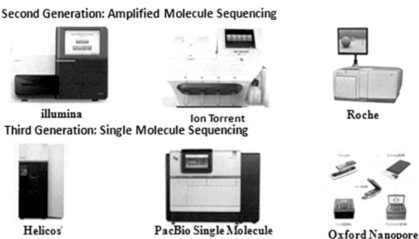

| illumina | Ion Torrent | Roche |

**Third Generation: Single Molecule Sequencing**

| Helicos | PacBio Single Molecule | Oxford Nanopore |

**Fig. 2** Generations of sequencing instrumentations

## 13 Illumina (Solexa) Sequencing

Illumina instruments are versatile and ideal for a large number of sequencing applications, including assembly, resequencing, transcriptome, SNP detection, and metagenomic studies (Tripathi et al., 2016). Illumina sequencing uses an elementary stage numerous techniques from the exemplary Sanger chain-termination technique. It makes use of sequencing by synthesis (SBS) technology: the massively parallel tracking of the addition of labeled nucleotides when copying the DNA strand (Slatko et al., 2018). Cutting-edge sequencing produces masses of DNA sequencing information and is more affordable and less tedious than customary Sanger sequencing. Illumina sequencing systems can have a data output range of 300, depending on the device type and configuration, with a capacity of kilobases to several terabases in a single run (Low & Tammi, 2016).

Solexa was a key player in the introduction of NGS, which Illumina procured. The vital improvement of the Illumina stage was "bridge reinforcement" that permits the arrangement of exceptional groups of intensified parts across a silicon chip (Anderson & Schrijver, 2010; Slatko et al., 2018; Amarasinghe et al., 2020). Over time, the number of clusters that could be read grew remarkably, and Illumina instruments became the first commercially accessible parallel sequencing technology. Other tools developed during the same period, such as the ion Torrent platform, became part of the NGS technology layout. NGS platforms are the leading type of sequencing technology used nowadays.

## 14   SOLiD Sequencing

The SOLiD line of instruments has a high throughput, allowing for a large number of brief scans to be generated. The SOLiD platform can be used for de novo sequencing, differential transcript expression, and resequencing (Pareek et al., 2011). The platform's short reads are a flaw that makes assembly difficult. SOLiD is an enzymatic sequencing method that employs DNA ligase, a biotech enzyme recognized for its capacity to ligate double-stranded DNA strands. Emulsion PCR is a way for immobilizing and amplifying an ssDNA primer-binding region (called an adaptor) on a bead that has been conjugated to the goal collection (i.e., the collection to be sequenced) (Xiao & Zhou, 2020). These beads are subsequently deposited onto a glass surface, resulting in a high density of beads, boosting the technique's throughput.

## 15   Ion Torrent Semiconductor Sequencing

The ion Torrent technology works by detecting the release of hydrogen ions as new nucleotides are incorporated into the growing DNA matrix (Barba et al., 2020; Gupta & Gupta, 2020). When the polymerase naturally incorporates nucleotides into the DNA strand, hydrogen ions are released as a by-product. Ion Torrent and its ion personal genome machine (PGM) sequencer use a high-density, micromachined array of holes for parallel nucleotide. Each well carries a different DNA template. Under the wells is an ion-sensitive layer followed by a proprietary ion sensor. Ions will change the pH of the solution detected by the ion sensor. If there are two identical bases in the DNA strand, the output voltage will be doubled, and the chip will be recorded on two identical bases, which is called no scan, camera, and light.

## 16   PacBio Sequencing

The PacBio RS/RS II thinks outside the box of other short peruses high-throughput sequencing instruments by zeroing in on length (Amarasinghe et al., 2020). The peruses, averaging ~4.6 kb, are altogether more than other sequencing stages making it ideal for sequencing short genomes like microbes or infections. Different benefits incorporate its capacity to arrangement areas of high G/C substance and decide the situation with changed bases (methylation, hydroxymethylation), killing the requirement for synthetic transformation during library readiness. The instrument's low yield of peruses keeps it valuable for getting together medium to huge genomes. Since quite a while ago, a different methodology read sequencing, utilizing pore-framing proteins and electrical identification was embraced by Oxford Nanopore Technologies (ONT). SMRT Sequencing has various benefits. Most outstanding, possibly, is its ability to create long reads. These long reads make it possible to span large structural variants and challenging repetitive regions that puzzle short-read sequencers because their short snippets cannot be differentiated from each other during assembly (Gong et al., 2018). Another advantage is low GC bias, allowing PacBio Systems to sequence through extreme-GC at AT regions that cannot be amplified during cluster generation on short-read platforms. A third advantage is detecting DNA methylations while sequencing since no amplification is done on the instrument. SMRT Sequencing, sometimes known as third-generation sequencing, had meticulous value for applications, including de novo genome sequencing, phasing, and detection of structural variants, epigenetic characterization, and sequencing of the transcriptome without the need for assembly (Ardui et al., 2018). Technology improvements over time improved the throughput and accurateness of SMRT Sequencing platforms. At present, SMRT Sequencing has industry-leading accuracy because of its HiFi sequencing. It is used worldwide to produce reference-grade genomes for microbes, plants, animals, and humans (Ardui et al., 2018).

# 17  The Basic Protocol for DNA Sequencing

1. Nucleic acid isolation is the first step in DNA sequencing from an isolate or clinical sample (Fig. 1). Any pathogen is likely to require troubleshooting to create a successful nucleic acid isolation protocol. Different standard kits and protocols can be used to test the presence of bacteria, mycobacteria, viruses, etc., in the sample. Different standard kits and protocols are required. While the protocols for optimal nucleic acid isolation may vary, the goal is to obtain high-quality, pure DNA. Contamination of the starting material can affect the quality of the sequencing data.
2. NGS requires the DNA to be of random fragments of similar size. These random fragments are fluorescently labeled for the sequencing process (Fig. 3). Library preparation is characteristically a straightforward process. However, optimization is necessary based on the laboratory instrumentation (e.g., thermal cycler) and laboratory conditions. The library prep kit selected can affect the quality of the sequencing information.
3. Sequencing is relatively straightforward and depends on the sequencer in the laboratory. No manual time is required once the sequencing runs have started. The duration of the sequencing runs depends on the sequencer, the number of samples, the number of reads, etc. (Fig. 3). More than an hour should be reserved for sequencing.

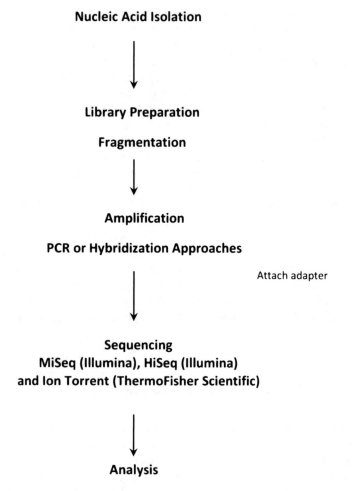

**Fig. 3** Basic protocol pipeline for DNA sequencing

4. During sequencer data analysis, the data are saved and ultimately analyzed.

## 18　Precautions

The following sequencing delivery precautions should be followed:

- Make sure vials and tips are free of DNase.
- Always wear gloves when handling the specimen, and ensure that your bench, pipettes, tips, and solutions are RNase-free.
- It is good to use the columns to purify the DNA/RNA. Columns remove the salts present in the DNA.

- Send samples with enough ice.
- Wipe up spilled solutions and dispose of them in suitable biohazardous containers.
- Be careful with the lids of PCR devices.
- Define and use separate areas for sample preparation, PCR setup, and post-PCR analysis.
- Restrict equipment to these areas. Keep the PCR machine and electrophoresis apparatus in the post-PCR area.
- Prepare and store PCR reagents separately and only use them for their intended purpose. Aliquot the reagents in small portions, and store them in any location according to their use in pre-PCR or post-PCR applications. Store aliquots separately from other DNA samples.
- Use separate sets of pipettes and pipette tips, lab coats, glove boxes, and trash cans for the pre- and post-PCR areas. When preparing the NGS library, clean the work surfaces and pipettes beforehand with a nucleic acid decontaminant.
- Use pipettes and pipette tips with special aerosol filters to prepare DNA samples and reaction mixtures.
- Follow the golden rule of PCR: Do not bring any reagents, equipment, or pipettes used in a post-PCR area into the pre-PCR area. This even applies to your lab notebook and pens. Label your pre- and post-PCR items so you can easily see where they belong.
- Keep the number of PCR cycles to a minimum, as highly sensitive assays are more prone to contamination effects.
- Do not use too much DNA in the sequencing reaction. You should try not to use more than 25 ng of the template times the template length in kilobases.
- Linearize high G + C plasmids before sequencing.

## 19  Need of DNA Sequencing Technology

(a) Sequencing technology is currently used to learn the complexities of genomic architecture in a wide variety of species, including humans, animals, plants, and microorganisms (Mitchelson, 2005; Liu et al., 2012; Suwinski et al., 2019) (Fig. 4).

(b) Sequencing offers conventional and rapid methods of disease diagnosis, prognosis, therapy decision-making, and patient follow-up. Validating potential drug targets with the ultimate goal of considering potential drug treatment options is essential in drug discovery. A large-scale approach to data generation using NGS and bioinformatic data analysis has identified new biomarkers expected to aid in diagnosing and treating diseases soon. Efficient analysis of substantial information databases by new computer tools can lead to the development of safer drugs with superior efficacy and lower toxicity, thereby improving patient health and preventing disease (Fig. 4).

(c) It offers new opportunities for personalized precision in medicine, as genomics, economics, metagenomics, epigenomics, epigenetics, and transcriptomics have

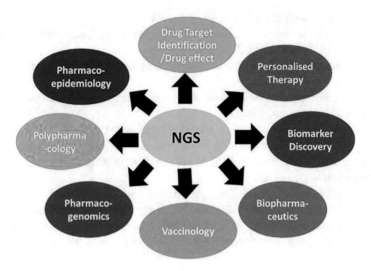

**Fig. 4** Applications of DNA sequencing in different biological fields

become a certainty (Mitchelson, 2005; Leamon & Rothberg, 2009; Suwinski et al., 2019). For example, epigenetic modification analysis can be performed using sequencing technologies to identify any major epigenetic changes in a patient caused by exposure to drugs or pathologies.

(d) Sequencing data can provide a clear understanding of the spectrum of antibodies, leading to engineered antibodies that specifically target the desired antigens in generalized vaccines (Suwinski et al., 2019).

## 20    Conclusion

The progress of sequencing technology in recent years has been impressive. The perfect sequencing platform can process a single DNA or RNA molecule with advantages such as no requirement of preliminary amplification, no need for optical steps, read lengths from MB to GB, no GC shift, high reading accuracy, and flexible enough to generate as many continuous readings as possible according to the needs of specific research questions. In addition, it should be cheap to buy and operate and easy to use, has short delivery time, and has a separate zero library creation; of course, such a sequencing platform does not yet exist.

## References

Alberts, B., Johnson, A., Lewis, J., Raff, M., Roberts, K., & Walter, P. (2002). Isolating, cloning, and sequencing DNA. Molecular Biology. Cell 4th Ed. Retrieved June 23, 2021, from https://www.ncbi.nlm.nih.gov/books/NBK26837/

Amarasinghe, S. L., Su, S., Dong, X., Zappia, L., Ritchie, M. E., & Gouil, Q. (2020). Opportunities and challenges in long-read sequencing data analysis. *Genome Biology, 21*, 30. https://doi.org/10.1186/s13059-020-1935-5

Anderson, M. W., & Schrijver, I. (2010). Next generation DNA sequencing and the future of genomic medicine. *Genes, 1*, 38–69. https://doi.org/10.3390/genes1010038

Ardui, S., Ameur, A., Vermeesch, J. R., & Hestand, M. S. (2018). Single-molecule real-time (SMRT) sequencing comes of age: Applications and utilities for medical diagnostics. *Nucleic Acids Research, 46*, 2159–2168. https://doi.org/10.1093/nar/gky066

Bansal, G., Narta, K., & Teltumbade, M. R. (2018). Next-generation sequencing: Technology, advancements, and applications. In A. Shanker (Ed.), *Bioinformatics: Sequences, structures, phylogeny* (pp. 15–46). Springer. https://doi.org/10.1007/978-981-13-1562-6_2

Barba, E., Tsermpini, E.-E., Patrinos, G. P., & Koromina, M. (2020). Chapter 8—Genome informatics pipelines and genome browsers. In G. P. Patrinos (Ed.), *Applied genomics and public health translational and applied genomics* (pp. 149–169). Academic Press. https://doi.org/10.1016/B978-0-12-813695-9.00008-X

Barba, M., Czosnek, H., & Hadidi, A. (2014). Historical perspective, development and applications of next-generation sequencing in plant virology. *Viruses, 6*, 106–136. https://doi.org/10.3390/v6010106

Berg, J. M., Tymoczko, J. L., & Stryer, L. (2002). DNA polymerases require a template and a primer. Biochemistry. 5th Ed. Retrieved June 23, 2021, from https://www.ncbi.nlm.nih.gov/books/NBK22374/

Buermans, H. P. J., & den Dunnen, J. T. (2014). Next-generation sequencing technology: Advances and applications. *Biochimica et Biophysica Acta—Molecular Basis of Disease, 1842*, 1932–1941. https://doi.org/10.1016/j.bbadis.2014.06.015

Choudhuri, S. (2014). Chapter 7—Additional bioinformatic analyses involving nucleic-acid sequences**the opinions expressed in this chapter are the author's own and they do not necessarily reflect the opinions of the FDA, the DHHS, or the Federal Government. In S. Choudhuri (Ed.), *Bioinformatics for beginners* (pp. 157–181). Academic Press. https://doi.org/10.1016/B978-0-12-410471-6.00007-4

Clark, D. P., & Pazdernik, N. J. (2013). Chapter 9—Genomics & Systems Biology. In D. P. Clark & N. J. Pazdernik (Eds.), *Molecular Biology* (2nd ed., pp. 248–272). Academic Press. https://doi.org/10.1016/B978-0-12-378594-7.00009-3

Clark, D. P., Pazdernik, N. J., & McGehee, M. R. (2019). Chapter 6—Polymerase Chain reaction. In D. P. Clark, N. J. Pazdernik, & M. R. McGehee (Eds.), *Molecular Biology* (3rd ed., pp. 168–198). Academic Cell. https://doi.org/10.1016/B978-0-12-813288-3.00006-9

Deharvengt, S. J., & Tsongalis, G. J. (2018). Chapter 31—Molecular assessment of human diseases in the clinical laboratory. In W. B. Coleman & G. J. Tsongalis (Eds.), *Molecular pathology* (2nd ed., pp. 709–730). Academic Press. https://doi.org/10.1016/B978-0-12-802761-5.00031-6

Delaney, C., Garg, S. K., & Yung, R. (2015). Analysis of DNA methylation by pyrosequencing. *Methods Mol. Biol. Clifton NJ, 1343*, 249–264. https://doi.org/10.1007/978-1-4939-2963-4_19

Diaz-Sanchez, S., Hanning, I., Pendleton, S., & D'Souza, D. (2013). Next-generation sequencing: The future of molecular genetics in poultry production and food safety1 1Presented as part of the next generation sequencing: Applications for food safety and poultry production symposium at the poultry science Association's annual meeting in Athens, Georgia, July 10, 2012. *Poultry Science, 92*, 562–572. https://doi.org/10.3382/ps.2012-02741

García-sancho, M. (2010). A new insight into Sanger's development of sequencing: From proteins to DNA, 1943-1977. *Journal of the History of Biology, 43*, 265–323.

Gaudin, M., & Desnues, C. (2018). Hybrid capture-based next generation sequencing and its application to human infectious diseases. *Frontiers in Microbiology, 9*. https://doi.org/10.3389/fmicb.2018.02924

Gong, L., Wong, C.-H., Cheng, W.-C., Tjong, H., Menghi, F., Ngan, C. Y., et al. (2018). Picky comprehensively detects high-resolution structural variants in nanopore long reads. *Nature Methods, 15*, 455–460. https://doi.org/10.1038/s41592-018-0002-6

Gupta, A. K., & Gupta, U. (2020). Chapter 20—Next-generation sequencing and its applications. In A. S. Verma & A. Singh (Eds.), *Animal biotechnology* (2nd ed., pp. 395–421). Academic Press. https://doi.org/10.1016/B978-0-12-811710-1.00018-5

Gupta, A. K., & Gupta, U. D. (2014). Chapter 19—Next generation sequencing and its applications. In A. S. Verma & A. Singh (Eds.), *Animal biotechnology* (pp. 345–367). Academic Press. https://doi.org/10.1016/B978-0-12-416002-6.00019-5

Head, S. R., Komori, H. K., LaMere, S. A., Whisenant, T., Van Nieuwerburgh, F., Salomon, D. R., et al. (2014). Library construction for next-generation sequencing: Overviews and challenges. *BioTechniques, 56*, 61–77. https://doi.org/10.2144/000114133

Heather, J. M., & Chain, B. (2016). The sequence of sequencers: The history of sequencing DNA. *Genomics, 107*, 1–8. https://doi.org/10.1016/j.ygeno.2015.11.003

Hegedüs, É., Kókai, E., Nánási, P., Imre, L., Halász, L., Jossé, R., et al. (2018). Endogenous single-strand DNA breaks at RNA polymerase II promoters in Saccharomyces cerevisiae. *Nucleic Acids Research, 46*, 10649–10668. https://doi.org/10.1093/nar/gky743

Hogan, K. (2006). Chapter 4—Principles and techniques of molecular biology. In H. C. Hemmings & P. M. Hopkins (Eds.), *Foundations of anesthesia* (2nd ed., pp. 51–69). Mosby. https://doi.org/10.1016/B978-0-323-03707-5.50010-3

Hughes, R. A., & Ellington, A. D. (2017). Synthetic DNA synthesis and assembly: Putting the synthetic in synthetic biology. *Cold Spring Harbor Perspectives in Biology, 9*, a023812. https://doi.org/10.1101/cshperspect.a023812

Jennings, L. J., Arcila, M. E., Corless, C., Kamel-Reid, S., Lubin, I. M., Pfeifer, J., et al. (2017). Guidelines for validation of next-generation sequencing–based oncology panels: A joint consensus recommendation of the Association for Molecular Pathology and College of American pathologists. *The Journal of Molecular Diagnostics, 19*, 341–365. https://doi.org/10.1016/j.jmoldx.2017.01.011

Kircher, M., & Kelso, J. (2010). High-throughput DNA sequencing—Concepts and limitations. *BioEssays, 32*, 524–536. https://doi.org/10.1002/bies.200900181

Leamon, J. H., & Rothberg, J. M. (2009). DNA sequencing and genomics. In M. Schaechter (Ed.), *Encyclopedia of microbiology* (3rd ed., pp. 148–161). Academic Press. https://doi.org/10.1016/B978-012373944-5.00024-9

Liu, L., Li, Y., Li, S., Hu, N., He, Y., Pong, R., et al. (2012). Comparison of next-generation sequencing systems. *Journal of Biomedicine & Biotechnology, 2012*, e251364. https://doi.org/10.1155/2012/251364

Low, L., & Tammi, M. T. (2016). Introduction to next generation sequencing technologies. *Bioinformatics (WORLD SCIENTIFIC)*, 1–21. https://doi.org/10.1142/9789813144750_0001

Maxam, A. M., & Gilbert, W. (1986). A method for determining DNA sequence by labeling the end of the molecule and cleaving at the base. Isolation of DNA fragments, end-labeling, cleavage, electrophoresis in polyacrylamide gel and analysis of results. *Molekuliarnaia Biologiia (Mosk), 20*, 581–638.

McCarty, C. A., Wilke, R. A., Giampietro, P. F., Wesbrook, S. D., & Caldwell, M. D. (2005). Marshfield clinic personalized medicine research project (PMRP): Design, methods and recruitment for a large population-based biobank. *Personalized Medicine, 2*, 49–79. https://doi.org/10.1517/17410541.2.1.49

Mitchelson, K. R. (2005). DNA SEQUENCING. In P. Worsfold, A. Townshend, & C. Poole (Eds.), *Encyclopedia of analytical science* (2nd ed., pp. 286–293). Elsevier. https://doi.org/10.1016/B0-12-369397-7/00683-X

Nimse, S. B., Song, K., Sonawane, M. D., Sayyed, D. R., & Kim, T. (2014). Immobilization techniques for microarray: Challenges and applications. *Sensors, 14*, 22208–22229. https://doi.org/10.3390/s141222208

Pareek, C. S., Smoczynski, R., & Tretyn, A. (2011). Sequencing technologies and genome sequencing. *Journal of Applied Genetics, 52*, 413–435. https://doi.org/10.1007/s13353-011-0057-x

Pasipoularides, A. (2017). Genomic translational research: Paving the way to individualized cardiac functional analyses and personalized cardiology. *International Journal of Cardiology, 230*, 384–401. https://doi.org/10.1016/j.ijcard.2016.12.097

Reuter, J. A., Spacek, D., & Snyder, M. P. (2015). High-throughput sequencing technologies. *Molecular Cell, 58*, 586–597. https://doi.org/10.1016/j.molcel.2015.05.004

Roden, D. M., Wilke, R. A., Kroemer, H. K., & Stein, C. M. (2011). Pharmacogenomics: The genetics of variable drug responses. *Circulation, 123*, 1661–1670. https://doi.org/10.1161/CIRCULATIONAHA.109.914820

Sanger, F., Nicklen, S., & Coulson, A. R. (1977). DNA sequencing with chain-terminating inhibitors. *Proceedings of the National Academy of Sciences of the United States of America, 74*, 5463–5467.

Shetty, P. J., Amirtharaj, F., & Shaik, N. A. (2019). Introduction to nucleic acid sequencing. In N. A. Shaik, K. R. Hakeem, B. Banaganapalli, & R. Elango (Eds.), *Essentials of bioinformatics, volume I: Understanding bioinformatics: Genes to proteins* (pp. 97–126). Springer International Publishing. https://doi.org/10.1007/978-3-030-02634-9_6

Shokralla, S., Spall, J. L., Gibson, J. F., & Hajibabaei, M. (2012). Next-generation sequencing technologies for environmental DNA research. *Molecular Ecology, 21*, 1794–1805. https://doi.org/10.1111/j.1365-294X.2012.05538.x

Slatko, B. E., Gardner, A. F., & Ausubel, F. M. (2018). Overview of next generation sequencing technologies. *Current Protocols in Molecular Biology, 122*, e59. https://doi.org/10.1002/cpmb.59

Suwinski, P., Ong, C., Ling, M. H. T., Poh, Y. M., Khan, A. M., & Ong, H. S. (2019). Advancing personalized medicine through the application of whole exome sequencing and big data analytics. *Frontiers in Genetics, 10*. https://doi.org/10.3389/fgene.2019.00049

Tripathi, R., Sharma, P., Chakraborty, P., & Varadwaj, P. K. (2016). Next-generation sequencing revolution through big data analytics. *Front. Life Sci., 9*, 119–149. https://doi.org/10.1080/21553769.2016.1178180

van Holde, K. E., & Zlatanova, J. (2018). Chapter 14—Recombinant DNA: The next revolution. In K. E. van Holde & J. Zlatanova (Eds.), *The evolution of molecular biology* (pp. 149–163). Academic Press. https://doi.org/10.1016/B978-0-12-812917-3.00014-0

Xiao, T., & Zhou, W. (2020). The third-generation sequencing: The advanced approach to genetic diseases. *Transl. Pediatr., 9*, 163–173. https://doi.org/10.21037/tp.2020.03.06

Xue, Y., Wang, Y., & Shen, H. (2016). Ray Wu, fifth business or father of DNA sequencing? *Protein & Cell, 7*, 467–470. https://doi.org/10.1007/s13238-016-0271-8

Zhao, S., Watrous, K., Zhang, C., & Zhang, B. (2017). Cloud computing for next-generation sequencing data analysis. *IntechOpen*. https://doi.org/10.5772/66732

Zhou, X., & Li, Y. (Eds.). (2015). Chapter 2—Techniques for oral microbiology. In *Atlas of oral microbiology* (pp. 15–40). Academic Press. https://doi.org/10.1016/B978-0-12-802234-4.00002-1

# Index

Printed in the United States
by Baker & Taylor Publisher Services